THE
CHILLING
STARS

A Cosmic View of
Climate Change

THE CHILLING STARS

A Cosmic View of Climate Change

**Henrik Svensmark
& Nigel Calder**

ICON BOOKS UK
TOTEM BOOKS USA

Originally published in 2007 by Icon Books Ltd

This updated edtion published in the UK in 2008
by Icon Books Ltd, The Old Dairy, Brook Road,
Thriplow, Cambridge SG8 7RG
email: info@iconbooks.co.uk
www.iconbooks.co.uk

Sold in the UK, Europe, South Africa and Asia
by Faber & Faber Ltd, 3 Queen Square, London WC1N 3AU
or their agents

Distributed in the UK, Europe, South Africa and Asia
by TBS Ltd, Frating Distribution Centre, Colchester Road
Frating Green, Colchester CO7 7DW

This edition published in Australia in 2008
by Allen & Unwin Pty Ltd, PO Box 8500,
83 Alexander Street, Crows Nest, NSW 2065

This edition published in the USA in 2008 by Totem Books
Inquiries to Icon Books Ltd, The Old Dairy, Brook Road,
Thriplow, Cambridge SG8 7RG, UK

Distributed in Canada by Penguin Books Canada,
90 Eglinton Avenue East, Suite 700, Toronto, Ontario M4P 2YE

ISBN: 978-1840468-66-3

Typesetting by Wayzgoose
Line drawings by HallidayBooks.com

Printed and bound in the UK by
Clays of Bungay

Contents

List of illustrations

Black and white images in the text

Colour plate section

Foreword by Eugene Parker

Over the past 50 years geological and geophysical studies of the fossil record of past climate have established the recurring extreme swings this old planet of ours has experienced during its long life. The climate has varied between Snowball Earth, frozen over from the poles to the Equator, to extended periods of warm climate extending all the way to the poles. These recurring climatic states are influenced by a variety of circumstances, such as the positions of the drifting continents and the associated ocean currents, the evolving composition of the atmosphere, the gradual evolutionary brightening of the Sun, and the precession of the spin axis of Earth combined with the variations of the eccentricity of the orbit. It has also been established that the contemporary climate responds strongly to some aspect of the magnetic activity of the Sun, including a small brightening of about one part in a thousand when the Sun is active.

And yet these effects do not add up to a scientific understanding. Too little is known about the ancient conditions, no specific driver has been identified, and the recent centuries of moderate swings between warm and cold, in association with solar magnetic activity, are driven much more strongly than the small brightening of the Sun can explain.

Given this impasse, it is our good fortune that, some years ago, Henrik Svensmark recognised the importance of cloud cover in the temperature control of planet Earth. Clouds are highly reflective to incoming sunlight. Svensmark also recognised that the individual water droplets that make up a cloud form mostly where ions have been created by passing cosmic ray particles, thereby tying cloud formation to the varying cosmic ray intensity. That is to say, cosmic rays control the powerful 'cloud valve' that regulates the heating of Earth. There is an immense task ahead to quantify the effect, with some degree of urgency in view of the present global warming.

Curiously, the urgency has not immediately facilitated the acceptance and support for this research that one might reasonably expect. Global warming has become a political issue both in governments and in the scientific community. The scientific lines have been drawn by 'eminent' scientists, and an important new idea is an unwelcome intruder. It upsets the established orthodoxy.

This is not an unfamiliar phenomenon. I recall when I was young and had demonstrated that the million-degree corona of the Sun expands to form the supersonic solar wind, thereby explaining the 'solar corpuscular radiation' responsible for the anti-solar pointing of comet tails and the variations of the cosmic rays. The hydrodynamic solar wind was an unwelcome idea and was accepted for publication in a prominent scientific journal only because the editor over-rode the heartfelt objections of the two 'eminent' referees.

Svensmark received harsher treatment for his scientific creativity, and he found it hard to achieve a secure position with adequate funding. He is in good company, when we recall that Jack Eddy lost his job when he confirmed and extended the earlier work of Walter Maunder, who had pointed out that the Sun showed a significant dearth of sunspots over the extended period 1645–1715. Eddy emphasised the important point that the Maunder Minimum was a period of cold terrestrial climate, thereby making the first direct connection of climate to solar magnetic activity.

Svensmark was not idle in spite of the persistent difficulties about funding his pursuit of the cosmic ray–cloud connection. He was busy in his Copenhagen laboratory with a relatively simple but affordable experiment that clearly demonstrated the essential role of ions in the air for the formation of clouds. Today a full-up experiment is being built at CERN in Geneva, starting with the fast particles from an accelerator to simulate the cosmic rays.

Fortunately, as soon as his Copenhagen experiment gave firm results, Svensmark joined forces with a science writer,

Nigel Calder, to explain the whole subject and its history in this simply worded book. As Svensmark's theory of cosmic rays, clouds, and climate penetrates a wide range of sciences, I suspect that specialists in various fields will profit from this plain-language introduction as much as general readers.

Eugene Parker

Half a century after Eugene Parker discovered the solar wind, he still works as Distinguished Service Professor Emeritus in physics, astronomy and astrophysics at the University of Chicago. His awards include the US National Medal of Science and the Kyoto Prize for Lifetime Achievement in Basic Science.

About the authors

Henrik Svensmark leads the Center for Sun-Climate Research at the Danish National Space Center. He previously held research positions at the University of California Berkeley, the Nordic Institute of Theoretical Physics, the Niels Bohr Institute and the Danish Meteorological Institute. He has published more than 50 scientific papers on theoretical and experimental physics, including nine landmark papers on climate physics. He received the Knud Højgaard Anniversary Research Prize in 1997 and the Energy-E2 Research Prize in 2001.

Nigel Calder has spent a lifetime spotting and explaining the big discoveries in all branches of science. He served his apprenticeship as a science writer on the original staff of the magazine *New Scientist*, and became its editor, 1962–66. Since then he has worked as an independent author and TV scriptwriter. He won the UNESCO Kalinga Prize for the Popularization of Science for his work for the BBC in a long succession of 'science specials', with accompanying books. His *Magic Universe* (OUP, 2003) was shortlisted for the Aventis Prize for Science Books 2004.

Authors' note

This book is the product of a year-long conversation. Intensive research was still in progress on several big questions, as the text evolved in 2005–06. While Svensmark supplied the core scientific input, Calder strung the words together and added various touches. Right up to the completion of the manuscript, we shared the thrill of finding out things that no one knew – including the results of urgent calculations done with the help of Svensmark's son Jacob.

We first met in 1996. Eigil Friis-Christensen introduced us over a lunch of Danish herring and lager, and Svensmark divulged his initial results showing that cosmic rays affect the Earth's cloud cover. Calder hurried off to write a book about the history of that discovery and its implications – *The Manic Sun* (Pilkington Press, 1997). Often in the years that followed we talked about a possible updated edition, until the story advanced in so many unexpected directions that only a new book would do.

We are grateful to those who made helpful comments on the draft manuscript as a whole or on parts of it. In alphabetical order they were Liz Calder, Peter Campbell, Roland Diehl, Jasper Kirkby, Gunther Korschinek, Eugene Parker, Jens Olaf Pepke Pedersen, Nir Shaviv and Ján Veizer. None of them is to blame for any faults that may remain.

Warm thanks are also due to Simon Flynn and his colleagues at Icon Books, for accepting the manuscript under strange conditions of confidentiality that preceded the announcement of an experimental result, and then producing the book with great speed and élan.

Henrik Svensmark *Hellerup, Copenhagen, Denmark*
Nigel Calder *Crawley, West Sussex, England*

Overview

On a clear, starry night you can catch a cold, and our ancestors were sometimes tempted to think that the Moon and the stars sucked heat from the Earth and made people ill. It was good observation but dodgy theorising. Astronomers will now tell you that most of the bright stars are far hotter than the Sun. Yet when the biggest of them expire in mighty supernova explosions they spray the Galaxy with atomic bullets, the charged particles known as cosmic rays. As a result, those exploded stars do indeed chill the world, by making it cloudier.

The discovery seemed crazy at first. Who would think that the ordinary clouds that decorate the sky take their orders from exploded stars far off in space? Or that the climate obeys the swarms of atomic particles that rain down on us from the Milky Way? But a recent experiment reveals how the trick is done, and thereby alters much of what scientists believed they knew about the weather, the climate, and the long history of life on the Earth.

To unwrap some of Nature's best-kept secrets, this book visits unlikely places, from the Atlantic sea floor to fossil-rich hills in China, and from the stormy Sun to the spiral arms of the Milky Way. Connections and surprises pop up from the chasms of space and time. As the book's wide scope might puzzle some readers, we offer this brief overview of the whole story at the outset. Skip it if you like.

The climate is always changing. The first clue that cosmic rays have something to do with it comes in Chapter 1, from the alternating episodes of warmth and cold over the past few thousand years. Most recently the Little Ice Age, which peaked around 300 years ago, has given way to the present warm interlude.

The Little Ice Age coincided with an unusual state of the Sun, known as the Maunder Minimum. Sunspots were very scarce, a symptom of feeble magnetic activity. A jump in the production rate of radiocarbon atoms and other long-lived tracers, made by cosmic rays in nuclear reactions in the air, was another indicator. We're shielded from many of the cosmic rays by the Sun's magnetic field, but when it weakens more of them can reach the Earth.

Chilling events like the Little Ice Age have occurred nine times since the most recent ice age ended 11,500 years ago, always associated with high counts of radiocarbon and other tracers. Historians and archaeologists testify to the misery caused to our ancestors. Reaching further back in time, into the ice ages, a German scientist found stones dropped onto the ocean floor by armadas of icebergs during a succession of very cold spells. Again, these coincided with low solar activity.

Among scientists who already agree that the Sun plays a prominent role in climate change, opinions differ about how it exerts its influence. Some want to explain the alternations

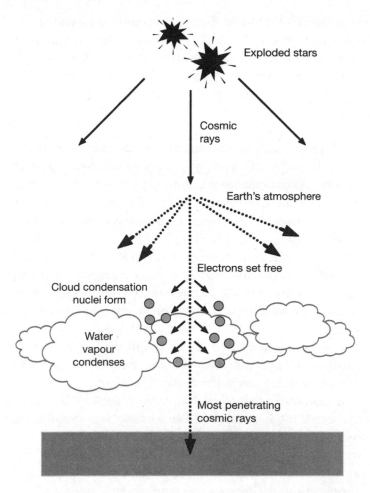

1. *The story in a nutshell: more cosmic rays mean more clouds and a chillier world – because the cosmic rays assist in cloud formation.*

of warm and chilly periods by changes in solar brightness. For them, the cosmic rays play no direct part in the weather but merely signal whether the Sun is more or less active, magnetically speaking, and therefore more radiant or dimmer. On the other hand, Danish scientists led by your author Svensmark think that a direct climatic role is more important, because cosmic rays affect the world's cloudiness.

Chapter 1 ends with an outline of the strongest evidence against the Svensmark theory, assembled by a Swiss physicist. About 40,000 years ago the Earth's magnetic field became very weak. Geophysicists call it the Laschamp Event. As a result, many more cosmic-ray particles entered the atmosphere, and left the atomic tracers of their passage. According to the theory of cosmic rays and clouds, shouldn't they have caused a severe cooling? But that didn't happen.

To rebut this well-reasoned argument, Svensmark looked again at the adventures of the cosmic rays, as related in Chapter 2. You don't notice it, but about twice a second a cosmic-ray particle whizzes through your head and disappears into the ground under your feet. When you climb a mountain, or fly in a jet plane, the rate is much higher.

Cosmic rays seemed like an optional extra, after an Austrian scientist detected them nearly a century ago. They were of great interest to scientists, certainly, but perhaps unimportant in the domestic economy of the Universe or the Earth. Only recently have astronomers realised that cosmic rays are an essential ingredient in the witch's brew from which come stars, planets and the chemicals needed for life. And in ways that the experts have been slow to appreciate, the cosmic rays arriving here from a distant chorus of exploded stars continue to influence our lives.

Before they can reach us, the cosmic rays must break through three defensive shields – the Sun's magnetism, the Earth's magnetism, and the air around us. Our planet's generous atmosphere is one reason why this planet is more congenial for life than the surface of Mars, where the cosmic rays are hundreds of times more intense. On the Earth, only the most energetic charged particles can travel right down to sea level. They are called muons, or heavy electrons, produced when the incoming cosmic rays hit the atmosphere.

In Svensmark's theory, the muons help to make clouds low in the air, which cool the world. To deal with the challenge of the Laschamp Event, he traced the origin of the muons, using a German computer program that calculates all the atomic and sub-atomic events that occur following the impacts of cosmic-ray particles on the air. He found that almost all the muons reaching the lowest 2,000 metres of the air are products of incoming particles too energetic to be affected by changes in the Earth's magnetism. So there's no reason to expect much of an increase in the muons in the Laschamp Event, or any significant cooling effect.

Do clouds just respond passively to climate changes due to other causes, as the mainstream climate science of the early 21st century has supposed? Or do they run the show? That is the theme of Chapter 3. Research in Copenhagen revealed what kinds of clouds are most important for climate change, and most influenced by the cosmic rays. They are low clouds that cover huge areas of the Earth – particularly noticeable on flights over the ocean, where they provide shiny but monotonous scenery for thousands of kilometres.

Unlike some of the higher clouds, which can have a warming effect, the clouds less than 3,000 metres above the surface keep the planet cool. When the penetrating cosmic rays are scarcer, the low clouds become fewer and the Earth grows warmer. During the 20th century, the Sun's magnetic shield more than doubled in strength and so reduced the cosmic rays and clouds enough to explain a large fraction of the global warming reported by climate scientists.

But are the clouds really in charge of climate change? Strong evidence that they are comes from the southern end of the Earth. Experts were bewildered by mounting evidence that Antarctica goes its own way. When the world as a whole warms up, Antarctica becomes colder, and vice versa. Complicated theories tried to account for this maverick behaviour. But if the clouds are in charge, this Antarctic climate anomaly is predictable. Antarctica is the only large region where clouds warm the snowy surface, while they chill the rest of the world.

Confirmation of cloud-driven climate change seems like good news for the world's inhabitants. It makes the Sun a potent agent of climate change, via its effects on cosmic rays, responsible for a substantial part of the warming in the 20th century. If so, the effect of carbon dioxide must be quite small and any global warming in the 21st century is likely to be much less than the typical predictions of 3 or 4 degrees Celsius.

For ten years after Svensmark and his colleagues in Copenhagen first pointed out the link between cosmic rays, clouds and climate, they were either ignored or criticised. Their idea could weaken the fashionable hypothesis about climate change, and the opposition was sometimes severe. It made research funding difficult to obtain. To face down the critics, and gain for their discovery the attention it deserved, the Danish team had to find out exactly *how* the cosmic rays influence cloud formation. Chapter 4 reveals the answer.

Strange to say, the experts on weather and climate never really knew where the clouds came from. Their elementary textbooks said that when humid air becomes cold enough, the moisture can condense to make clouds. But first there have to be small specks floating in the air, the cloud condensation nuclei on which the water droplets can form. The most important specks are themselves droplets, made from molecules of sulphuric acid and water. They need to be seeded too, and how that happened was a mystery. A research aircraft flying over the Pacific Ocean in 1996 discovered high-speed speck-making that contradicted all of the weathermen's prevailing theories.

The solution came from a big box of air in the basement of the Danish National Space Center, in 2005, in the experiment called SKY. Cosmic rays entering through the laboratory ceiling released electrons in the air, which then encouraged the clumping of molecules to make micro-specks, capable of gathering into the larger specks needed for cloud formation. The electrons' speed and efficiency in doing their work took the experimenters by surprise.

Other laboratory tests of the possible effects of cosmic rays in the atmosphere began in 2006. A multinational team collaborating at CERN, Europe's particle-physics lab in Geneva, was preparing the CLOUD experiment, more elaborate than SKY, and using accelerated particles to simulate the cosmic rays. The first run was with a replica of SKY, fitted with additional instruments.

The Copenhagen experiment completed the chain of explanation, from the cosmic rays released from exploding stars and battling through to the Earth's lower atmosphere, to their effects on the clouds and the climate. Scientists can now look much more confidently for the effects of cosmic-ray variations since the world began. As Chapter 5 explains, the influx of cosmic rays depends not just on the state of the Sun but on where we are in the Galaxy.

With the Earth in company, the Sun cruises among the stars, in an orbit around the centre of the Milky Way. Sometimes it finds itself in a dark region where hot, bright, explosive stars are few. There, cosmic rays are relatively scarce and the Earth's climate is warm. Geologists call it the hothouse mode. In other periods, when starlight and the cosmic rays are intense, the world goes into an icehouse phase, with glaciers and ice sheets forming part of the scenery.

An Israeli scientist adopted the Danish ideas about cosmic rays and climate to account for those major changes by visits to the bright spiral arms of the Milky Way. During the 500-million-year history of animal life on the Earth, switches between hothouse and icehouse have occurred four times. The cosmic-ray theory required an icehouse for the dinosaurs, because the Sun passed through a spiral arm during their tenure of the planet in the Mesozoic Era. Most geologists and fossil-hunters thought the era had been generally warm, but now firm evidence of ice on land comes from Australia. In those chilly times when cosmic rays were intense, small dinosaurs grew feathers to keep warm. As Chinese fossil-hunters have confirmed, some of them evolved into birds.

During its travels, the Sun also rises and plunges like a playful dolphin, up and down through the flat disc of the Galaxy, where the cosmic rays are most intense locally. This movement produces variations in climate about four times more frequent than those due to encounters with the spiral arms. As a sign that the cosmic-ray theory really works, that climatic record can now be used to refine the astronomers' facts and figures about the Milky Way.

Over billions of years the Galaxy itself changes, and events in the sky have sometimes provoked conditions so cold that glaciers and icebergs abound even in the tropics. Chapter 6 starts with this awesome state of affairs, which geologists call Snowball Earth. It happened in episodes around 2,300 and 700 million years ago.

These snowball events coincided with starbursts – frenzies of star-birth and star-death in the Milky Way provoked by a brush with another galaxy. With cosmic rays far more abundant than usual, and clouds making the world very gloomy, it froze over. Urgent adaptations of life led to great evolutionary changes. They included, in the last snowball events, the origin of animals.

On the other hand, the Earth was warmer than you might expect, early in its existence, when the young Sun was fainter than it is now. But the Sun was also much better at fending off the cosmic rays. That helped to create benign conditions for the earliest life, as discovered in rocks in Greenland, 3,800 million years old. Since then, living things have endured an ever-changing climate. The latest overview of the history of life shows intense cosmic rays prompting extraordinary lurches between scarcity and abundance.

During the past 3 million years some clusters of hot, explosive stars have ambushed the Sun and the Earth with a succession of nearby supernovae, which intensified the cosmic rays. Chapter 7 looks at possible links between those stellar cataclysms and the drying out of Africa that provoked the appearance of the first stone tools and the debut of human beings. At least one supernova was close enough to scatter exotic kinds of atoms on our planet, now retrieved from the ocean floor.

Several stars must have exploded in our cosmic neighbourhood and several sharp coolings occurred on the Earth. To try to match these events, in the hope of confirming cause and effect, is an exacting task for the latest gamma-ray detectors in orbit. The notion that human beings may owe their existence to those supernovae gives the astronomers a strong motive for their hunt. And this inquiry illustrates the extraordinary links between different branches of science that arise from the theory of cosmic rays, clouds and climate.

Cosmoclimatology, as we call it, emerges as a new field of science that opens up many opportunities for researchers. Chapter 8 sketches some of them, at the cutting edge of discovery in several areas. Knowledge of the Galaxy and the long histories of climate change and life on the Earth leaves vast room for improvement. And the new awareness of our special relationship to the Sun and its magnetic shield may help to narrow down the choice of places to search for alien life.

Meanwhile the Sun goes on regulating the influx of cosmic rays, but no one knows what it will do next. The real effects of human activity have to be reassessed. For these reasons, grand forecasts of climate are not to be trusted, but cosmoclimatology can offer practical advice to people exposed to hardship from climate change.

1 A lazy Sun launches iceberg armadas

Our ancestors endured shocking variations in climate · Events often matched changes in the Sun's behaviour · Rare atoms made by cosmic rays signal those changes · When their production increased, the world was chilled · But are the cosmic rays the agent, or merely a symptom?

A less public-spirited finder might have put the oddity up for sale on eBay, so the archaeologists of Bern Canton were grateful when Ursula Leuenberger presented them with an archer's quiver made of birch bark. They were amazed when radiocarbon dating showed the quiver to be 4,700 years old. Frau Leuenberger had picked it up while walking with her husband in the mountains above Thun. There, the perennial ice in the Schnidejoch had retreated in the unusually hot summer of 2003, revealing the relic hidden beneath it.

The hiking couple had unwittingly rediscovered a long-forgotten short-cut for travellers and traders across the barrier of the Swiss Alps. To keep treasure-hunters away, the find remained a secret for two years while archaeologists scoured the area of the melt-back and analysed the finds. By the end of 2005 they had some 300 items – from the Neolithic Era, the Bronze Age, the Roman period and medieval times.

The various ages of the items clustered in intervals when the pass of Schnidejoch was open, offering a quick route to and from the Rhone valley south of the mountains. There were no substantial human remains to compare with the murdered Ötztal 'ice man', found with a similar quiver high in the Italian Tyrol in 1991 and dated to 3300 BC. But the emergent history of repeated openings and closures of Schnidejoch gave a far more interesting picture of climate change.

The Ötztal man is a prize exhibit for those who assert that the climate at the start of the 21st century is alarmingly warm. The ice that preserved his mummified corpse lay unmelted, 3,250 metres above sea level, for more than 5,000 years – since the world was in its warmest phase following the most recent ice age. Then, so the story goes, the man-made global warming of the industrial era outstripped all natural variations and released the body as a warning to us all.

Quite different is the impression given by the relics found in the pass of Schnidejoch, at an altitude 500 metres lower than the Ötztal man's ice-tomb. They tell of repeated alternations between warm periods when the pass was useable and cold periods when it was shut by the ice. The discoveries also cleared up a long-standing mystery about a Roman lodging house found on the slopes above the present-day town of Thun, where there was a Roman temple and settlement. The head of the cantonal archaeological service, Peter Suter, explained his satisfaction at the outcome: 'We always asked ourselves why the lodging house was there. Now we know that it was on the route leading across the Schnidejoch.'

The youngest item found by the archaeologists was part of a shoe dating from the 14th or 15th century AD. It corresponds with the end of an interval known as the Medieval Warm Period. Thereafter the Schnidejoch was blocked by the glaciers of the Little Ice Age, the most recent period of intense cold. Nominally the Little Ice Age ended around 1850, but the gradual retreat of the ice took a century and a half to clear the pass, until its rediscovery early in the 21st century.

Here is a tale of natural variations in climate having a practical influence on the lives and travels of Europeans over 5,000 years. The climate was particularly cold in two periods around 800 BC and 1700 AD. Effects of the latter episode, the Little Ice Age, persisted in the Schnidejoch for so long that even the locals forgot that a useful pass was ever there.

The Medieval Warm Period and the Little Ice Age were an embarrassment for those who, in recent years, wished to play down the natural variations in climate that occurred before the Industrial Revolution. A widely publicised but now discredited graph of temperatures, produced in 1998 by Michael Mann of the University of Massachusetts and his colleagues, tried to iron out the variations. Lampooned as the hockey stick, Mann's graph showed the world remaining almost uniformly cool through most of the past 1,000 years until 1800. Then temperatures began to climb towards unprecedented highs in the late 20th century – so making the toe of the hockey stick and the supposed onset of an unprecedented episode of man-made global warming.

The relics from the Schnidejoch mock this Orwellian effort to make real-life events that were not politically correct disappear from climate history. They show that warming spells very like that of the past 100 years occurred repeatedly, long before the large-scale use of fossil fuels and the associated emissions of carbon dioxide gas were a possible factor. Attempts to argue that such events were not global are contradicted by abundant evidence for the Medieval Warm Period and the Little Ice Age from East Asia, Australasia, South America and South Africa, as well as from North America and Europe. Probing the errors that generated the hockey stick can be safely left to the statistical pathologists, while we explore the character and rhythms of climate change over centuries and millennia.

Sunspots missing in the Little Ice Age

Atomic bullets raining down from exploded stars, the cosmic rays, leave behind them business cards that record their split-second visits to the Earth's atmosphere. They take the form of unusual atoms created by nuclear reactions in the upper air. Especially valued by archaeologists as an aid to dating objects is radiocarbon, or carbon-14, made from nitrogen in the air.

Taken up into carbon dioxide, the gas of life by which plants grow, the carbon-14 finds its way via the plants and animals into wood, charcoal, bones, leather and other relics. The initial carbon-14 content corresponds to the amount prevailing in the air at the time of death. Then, over thousands of years, the atoms gradually decay back into nitrogen. If you see how much carbon-14 is left in an old piece of wood or fibre or bone, you can tell how many centuries or millennia have elapsed since the plant or animal was alive.

There's a snag about this gift from the stars, as archaeologists soon discovered. Some of their early radiocarbon dates seemed nonsensical, even contradictory – for example, a pharaoh of Egypt dated as being younger than his known successors. Hessel de Vries of Gronigen found the explanation in 1958. The rate of production of carbon-14 varies. Measurements in well-dated annual rings of growth in ancient trees sorted out the problem, and the archaeologists had more reliable, though often ambiguous dates.

And physicists could see changes over thousands of years in the performance of the Sun, as the chief gatekeeper of the cosmic rays. Its magnetic field protects us by repelling many of the cosmic rays coming from the Galaxy, before they can reach the Earth's vicinity. The variations that confused the archaeologists followed changes in the Sun's mood. Low production rates of carbon-14 meant that the Sun was very active, magnetically speaking. When it was lazy, more cosmic rays reached the Earth and the production of carbon-14 shot up.

The discovery opened the way to modern interpretations of the link between the Sun and the Earth's ever-changing climate, beginning in the 1960s. Roger Bray of New Zealand's Department of Scientific and Industrial Research traced the variations in the Sun's activity since 527 BC. He was able to connect increased production of radiocarbon by cosmic rays to other symptoms of feeble solar magnetic activity.

A scarcity of dark spots on the face of the Sun, which are made by pools of intense magnetism, was one such sign. Reports of auroras, which light the northern skies when the Sun is restless, were also scanty when the cosmic rays were making lots of radiocarbon. And most significantly, Bray linked solar laziness and high cosmic rays with historically recorded advances of glaciers, pushing their cold snouts down many valleys. The advances were most numerous in the 17th and 18th centuries, which straddled the coldest period of the Little Ice Age.

Some scientists are better than others in securing publicity for their work, and Bray was eclipsed a decade later by Jack Eddy of the High Altitude Observatory in Colorado. He had a snappier title for the peculiar state of the Sun in the late 17th century – the Maunder Minimum. In a report in 1976, Eddy so named it after Walter Maunder, superintendent of solar observations at London's Greenwich Observatory. In the 1890s Maunder wrote retrospectively about a period between 1645 and 1715 when the Sun was almost completely devoid of sunspots.

Vivid terms like Big Bang and black hole play an important part in the spread of scientific ideas, and Eddy was aware that he was on a winner with the Maunder Minimum.

> I knew I had a lot of selling to do if people were to accept the notion of such irregularity in the Sun, and I sought a name that people would remember. Maunder Minimum with all those m's had a kind of onomatopoeia. I think I did quite a bit for Maunder with that name, particularly because he got the idea from [Gustav] Sporer who was a German astronomer. So among the shots I took after publishing the paper were some from Germany that said, 'You know, you really named it after the wrong person' – which I knew very well.

In compensation, the name Sporer Minimum was later attached to another spell of low solar activity and high cosmic rays, between 1450 and 1540. Lesser events of a similar kind, 1300–60 and 1790–1820, are called the Wolf and Dalton Minima. The occurrence of four different episodes of solar enfeeblement, separated by brief periods of recovery, explains why historians of climate have often disagreed about when the Little Ice Age started and finished. But its severity is well documented, with glaciers bulldozing farms and villages, painfully brief summers, and death by famine in many places.

The violin-maker Antonio Stradivari lived during the Maunder Minimum, when the trees of Europe were faring badly, with the narrowest rings of annual growth seen in the past 500 years. That may explain why a Stradivarius can now be worth $10 million or more. In 2003 a tree-ring expert, Henri Grissino-Mayer of the University of Tennessee, and a climate scientist, Lloyd Burckle of Columbia University, pointed out that the narrow rings made the spruce wood used by Stradivari exceptionally strong and dense, with musical qualities that later violin-makers have never been able to match.

Dissent among the Sun's fans

Astrophysicists as well as climate scientists became fascinated by the Maunder Minimum. Sun-like stars, observed routinely for a quarter of a century or more, have revealed their capacity to shut down magnetic activity, just as the Sun did 300 years ago. In 1993, Robert Jastrow of Mount Wilson Observatory in California and Sallie Baliunas of the Harvard-Smithsonian Center for Astrophysics reported that, among twelve sun-like stars, most showed cycles of activity similar to the Sun's, while one of them, Tau Ceti, was virtually inert magnetically. Most dramatically, another star, 54 Piscium, behaved normally until 1980 when its magnetic activity suddenly slumped and remained low, as if it had just entered a Maunder-like minimum.

The man who first suggested to Jack Eddy that he

should look into Walter Maunder's story of the missing sunspots of the Little Ice Age was Eugene Parker. This Chicago physicist authored the theory of the solar wind by which the Sun creates its magnetic barrier against the cosmic rays. And at a conference in Tenerife in 2000, Parker called for much more attention to be paid to such variations in other stars.

> We know from observations of a few sun-like stars that one of them lost 0.4 per cent of its luminosity in only a few years. If the Sun did that, it could quickly reproduce the cold conditions of 300 years ago when solar activity was much reduced in the interval we call the Maunder Minimum. To find out what the Sun might do one day, we should set up an automated system to watch a thousand sun-like stars.

Like many other supporters of the Sun's role in climate change, including Jack Eddy and Sallie Baliunas, Parker assumed that the cold of the Little Ice Age was due to a fainter Sun, when the count of sunspots was low. For these scientists, variations in the intensity of visible and/or invisible rays coming from the Sun were what affected the climate. The increase in cosmic rays during the Little Ice Age was, in their view, merely a symptom of solar weakness and not itself a cause of the cooling.

In contrast, the authors of this book regard the decrease in the Sun's luminosity as only a contributory factor in the Little Ice Age and other such events. It was in Copenhagen in 1996 that Svensmark first noticed that a greater influx of cosmic rays brings an increase in global cloudiness, which can cause a much stronger cooling. But he then had to battle on two fronts, against those who thought the Sun had little influence in any case, and others who considered the solar contribution to be very important but dismissed the role of cosmic rays in climate change. The persistent dispute did not prevent the general facts about Sun–climate connections becoming ever more convincing.

Ice-rafting events

Victims of the Little Ice Age included the Viking settlers in Iceland and Greenland, where ice encumbered the coasts. Famines afflicted the Icelanders, and in the case of Greenland all of the incomers disappeared, although the native Greenlanders survived. As ice spread southwards in the Atlantic Ocean and melted, it released alien grit that is still detectable in the sea floor.

Far below the stormy waves, the bottom-dwelling organisms and a slow rain of microscopic shells and other debris from the ocean surface spend millions of years silently building layer upon layer of sediment. Insert long tubes into the sea floor, from a research ship far above, and you can recover cores that show layers of different colours and compositions. Expert eyes read the layers like the pages of a history book going back in time – the deeper in the core, the older the events.

Among the rather plain white deposits of small shells typical of warm conditions, bands of silty and sandy material show up when the world is cold. Floating ice can raft this grit from far away and drop it when the ice melts. The ice-rafted detritus in the North Atlantic floor reveals that the Little Ice Age was just the latest of a long series of similar but often more severe cooling events occurring at intervals of about 1,500 years.

Preceding the Maunder Minimum was the miserable period around 800 BC, at the transition from Bronze Age to Iron Age. It was one of those times when the pass was closed at the Schnidejoch. In the Netherlands, an archaeo-logical site in West Friesland has revealed effects of prolonged cold and wet weather, with a rising water table that drove the inhabitants from their low-lying settlements and farms. It was also a time when the Sun was idling. The palaeo-ecologist Bas van Geel of the University of Amsterdam drew the event to the attention of Svensmark and Calder in 1997, and commented on the cause.

> This abrupt climate change occurred simultaneously with a sharp rise in radiocarbon starting around 850 BC and peaking around 760 BC. Just because no one knows for sure how the solar changes work on the climate is no excuse for denying their effect.

While the Friesland folk suffered, ice-rafted debris on the North Atlantic sea floor exceeded the deposits seen during the Maunder Minimum, but it was not nearly as pro-nounced as in some earlier episodes. Our story now takes us back in time, when these abrupt but relatively short-lived changes of climate were superimposed on the gradual changes that cause the ice to advance and retreat in major ice ages, which typically occur every 100,000 years. According to a theory popular since the 1970s, wobbles of the Earth in orbit seem to help to set the slow rhythms of the ice ages. The briefer events are due to the variations in the Sun that affect the cosmic rays.

Wild climatic excursions first became obvious in the 1980s when Hartmut Heinrich of the German Hydro-graphical Institute in Hamburg examined sea-bed cores from the European side of the North Atlantic. In sediments laid down during the most recent ice age, which ended 11,500 years ago, he found eleven distinct layers enriched in quartz sand. In six cases, there were rock fragments originating far away, with the first of these exceptional ice-rafting events occurring 60,000 years ago, and the last 17,000 years ago.

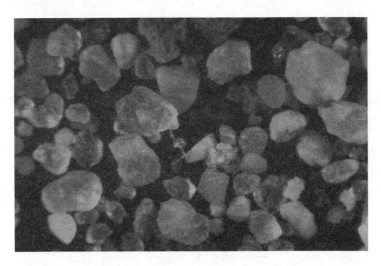

2. Grit from many places dropped by icebergs on the Atlantic floor about 22,000 years ago tells of one of the drastic chills known as ice-rafting events – in this case, Heinrich 2. All seem to have been linked to a lazy Sun and a high influx of cosmic rays. (Anne Jennings, Institute of Arctic and Alpine Research, University of Colorado-Boulder)

A Swiss student, Rüdiger Jantschik, traced the fragments to their sources. Those coming from the Norwegian Sea and Greenland were unsurprising, but there were also grains of white carbonate rock from northern Canada. Heinrich was amazed by the implications.

We had to imagine that, every so often, great armadas of icebergs broke out onto the sea. They could travel right across the Atlantic from North America, before depositing their debris on our side of the ocean. And they represented episodes of a kind not to be found in any of the text-books on climate change up to that time.

During the Heinrich Events, sudden drops of several degrees Celsius in mean temperatures could occur in the northern North Atlantic in the course of a human lifetime. The effects were felt far away. For example, in the Near East, recent investigations have shown that the level of Lake Lisan, at the site of the present Dead Sea, fell greatly during Heinrich Events, implying that there was much less rainfall there.

As they were arguably the most drastic changes in climate to which our ancestors were vulnerable – and by implication our descendants too – the Heinrich Events plainly deserved closer scrutiny. The world's largest collection of sea-bed cores is at Columbia University's Lamont-Doherty Earth Observatory, and there, in 1995, the geologist Gerard Bond set about examining North Atlantic cores more thoroughly.

Millimetre by millimetre, Bond searched for alien grains in the sea-bed deposits. The Heinrich Events became even more remarkable. Mixed with the white carbonates from the Hudson Strait region of northern Canada, Bond found grains stained red with haematite and originating from the St Lawrence region of southern Canada. Volcanic glass, black and translucent, came from Iceland. So the icebergs that rafted the material emanated simultaneously from widely scattered places. Some icebergs travelled as far as north-west Africa before melting and shedding their loads of grit.

Working with his wife Rusty Lotti, Bond found many more ice-rafting events during the most recent ice age than Heinrich had reported from the north-east Atlantic. They were easier to see in other parts of the ocean floor. The St Lawrence red and Iceland black grains sometimes showed up when the conspicuous white grains from the Hudson Strait were missing. The Heinrich Events were just the worst events in an often-repeated cycle.

Bond also carried the story of ice-rafting events forward in time, to the end of the ice age and into the generally warmer period since then. As geologists already knew, the great warming that terminated the ice age was interrupted about 13,000 years ago by a severe re-chilling called the Younger Dryas. At that time, the sea-floor cores showed an ice-rafting event of the Heinrich type – white grains and all.

After the ice age, the events left less conspicuous deposits – mainly of dust blown from northern lands and islands onto drifting sea-ice and transported southwards. But the deposits remained persistent in their rhythm, and several of them showed more ice-rafted debris than in the Little Ice Age. Besides the 800 BC event already mentioned, the sea floor reliably logs other cooling events well known to geologists and archaeologists: for example, around 8,300 years ago (6300 BC) and in the period 3600–3300 BC. The North Atlantic cooling episodes were linked with reduced rainfall at lower latitudes. If you're interested in the possible influences of climate change on human affairs, be aware that messages in clay envelopes dating from that latter cool period in Mesopotamia are the earliest known tax demands.

Widespread effects accompanied a cooling event that peaked at 1300 BC. While the sea-ice was shedding its dusty debris in the Atlantic, drought afflicted the countries around the eastern Mediterranean. The urban cultures of the Mycenaeans in Greece and the Hittites in Anatolia both collapsed. The Jews made their exodus from Egypt at a time when the Nile was running low. Disruption of the tin trade by robbers and pirates prompted experiments with iron and steel, notably in Cyprus, as a replacement for bronze.

The manic-depressive Sun

The coolings all coincided with lazy intervals on the Sun and increased cosmic rays, just as in the Maunder Minimum and the accompanying Little Ice Age. This was entirely unsurprising for those like Svensmark who expected such a link. Although he and others, including Bas van Geel in Amsterdam, pointed it out repeatedly, they were largely ignored. When Gerard Bond of Columbia first traced the milder versions of the Heinrich Events during the past 12,000 years, he was sceptical about the solar connection until Jürg Beer joined his team from the Swiss Federal Institute of Environmental Science and Technology.

Beer is an expert on past solar behaviour revealed by variations in radioactive beryllium-10, made in the atmosphere by cosmic rays. It has a much longer lifespan than carbon-14, and it does not become mixed up with living things or with the complicated cycles of carbon dioxide in the atmosphere and ocean. Settling atom by atom on the ice of Antarctica and Greenland, and buried there by later snowfalls, beryllium-10 provides an exceptionally valuable guide to solar behaviour over hundreds of thousands of years.

Thanks to heroic drilling projects at the coldest ends of the Earth, and the patient examination of the ice back home in the laboratories, long chronicles of climate change and its possible causes are now available. Besides the counts of beryllium-10, the ice layers store records of changing temperatures, along with traces of volcanic eruptions and of gases such as carbon dioxide and methane.

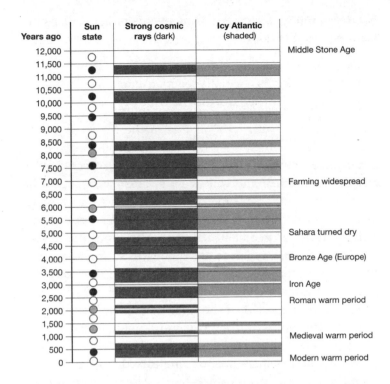

3. *Repeatedly during the past 12,000 years the Sun has weakened, letting in more cosmic rays from the Galaxy. The result was a chilly world, recorded in grit dropped by ice in the Atlantic – most recently in the Little Ice Age. The Modern Warm Period (often called global warming) is just the latest of a long succession of mild intervals when the Sun was more active and cosmic rays were relatively scarce. (Data from G. Bond and team, 2001)*

Although Beer did not favour Svensmark's climate-changing mechanism of cosmic rays and clouds, he was not at all averse to the proposition that the Sun somehow affects the climate in remarkable ways. When he used the cosmic rays as indicators of variable moods of the Sun, peaks in the beryllium-10 production seen in Greenland ice matched

Bond's carefully dated ice-rafting events rather well, as the team reported in 2001.

> Our correlations are evidence, therefore, that over the last 12,000 years virtually every centennial time scale increase in drift ice documented in our North Atlantic records was tied to a distinct interval of variable and, overall, reduced solar output.

Interspersed between the abrupt coolings of the Heinrich and Bond ice-rafting events were sudden warmings, discovered by Willi Dansgaard of Copenhagen and Hans Oeschger of Bern, when scrutinising ice cores retrieved by drilling into the Greenland ice sheet. Variations in the proportion of heavy oxygen atoms in the ice are an indicator of temperature changes. At two different drilling sites far apart on the ice sheet, in layers formed 45,000 to 15,000 years ago in the midst of the last ice age, the investigators saw a dozen sudden, strong warming events, each lasting a few hundred years.

More recent warm episodes repeatedly re-opened the Schnidejoch short-cut across the Alps. These were relatively mild, muted versions of the extravagant jumps in temperature during the ice age – just as the Little Ice Age was fortunately much less severe than a Heinrich Event. Is the Earth at present less vulnerable to drastic climate change than it was during the last ice age?

The two most recent upswings in temperature were the Medieval Warm Period and the global warming of the 20th century. Between about 1000 and 1300 AD many parts of the world were as warm if not warmer than now. Memorably, it was the heyday of the Vikings in the North Atlantic, the zenith of Muslim imperialism, culture and science, and in China a period so benign that the population doubled in

100 years. Europe's prosperity at that time is recorded in a boom in cathedral-building.

Vigorous activity in the Sun that curbed the influx of cosmic rays is plainly linked both to the Medieval Warm Period and to the 20th-century warming. The contrast with the high cosmic rays of the Little Ice Age gives a striking impression of the mood swings of the Sun. As Bond and his colleagues showed, the reduced levels of ice-rafting in the Atlantic seen in the Medieval Warm Period occurred altogether eight times during the past 12,000 years, whenever the cosmic-ray influx was low – with cold episodes in between. The similarity to the more drastic alternations of climate in the most recent ice age, between cold Heinrich Events and warm Dansgaard-Oeschger Events, leaves no doubt that solar variations were in charge of events in both periods.

Good times and bad, in the ice age

Modern human beings, spreading from Africa, made their first venture into western Europe during a Dansgaard-Oeschger warming around 35,000 years ago. Those Cro-Magnon people soon replaced the Neanderthal incumbents. No doubt the warmer conditions first lured them north and west into Europe. But their descendants were then on a climatic roller-coaster, with half a dozen big chills and re-warmings before the ice age ended. Nor was it only in Eurasia where courage and intelligence were tested to the limit, because drastic changes in rainfall patterns affected every inhabited part of the globe.

The Younger Dryas episode 13,000 years ago may have been especially irksome. As mentioned earlier, it came just when the ice age seemed to be ending. A peak in radio-carbon counts indicates that many more cosmic rays than usual were arriving when the cold conditions returned. Ice-rafted debris settled in abundance on the Atlantic sea-bed and reactivated glaciers crushed the forests that had been enticed up the valleys by warmer conditions.

A progressive increase in rainfall in Africa came to an abrupt end in the Younger Dryas and drought afflicted many places, with lake levels falling. And at Abu Hureyra, beside the River Euphrates in Syria, the inhabitants hit upon a new way of coping with climate change. That was where Gordon Hillman of London's Institute of Archaeology and his colleagues found evidence for the most consequential innovation in the whole of prehistory – the systematic cultivation of previously wild cereals, initially rye and wheat.

> The primary trigger appears to have been the critically reduced availability of key wild plant staples during the arid conditions of the Younger Dryas climatic episode. This early inception of cultivation then set the scene for the development and rapid spread of integrated agro-pastoral economies.

Chalk up a big human consequence of a surge in cosmic rays. Other such influences of climate change are now ripe for investigation. During the ice age the modern human beings gradually spread as far as Australia and Siberia, and eventually they crossed into North America. How did their journeys dovetail with the ever-changing climatic conditions – Dansgaard-Oeschger versus Heinrich – in the various parts of the world?

Before that great dispersal, the climate had taken its first deep plunge towards the coldest conditions of the ice age more than 70,000 years ago. Experts have raised particular questions about what happened, physically and biologically. Did the dust from a stupendous volcanic explosion at Toba in Sumatra, 74,500 years ago, darken and chill the world in a volcanic winter? Was the human population reduced so much as to produce a genetic bottleneck, such that all of us are descended from a few survivors of that catastrophe?

Captivating though these propositions are, the evidence is debatable. If human beings were nearly wiped out, many other species should have suffered too, and there is little sign of that. As for Toba's climatic impact, the violence of the event scattered ash as far as India and must have put huge quantities of dust into the stratosphere. But geologists from Taiwan have traced a previous super-eruption of Toba, 790,000 years ago, with about half the force of the later event. It was followed by a warming, in a transition from ice-age temperatures to interglacial conditions. Perhaps there was a brief cooling so far undetected, but no long-term effect.

Evidence about Toba comes from drilling into the ice sheets of Greenland and Antarctica. Indicators of prevailing temperatures, when ancient layers of ice were formed from fallen snow, are the varying counts of heavy oxygen atoms, oxygen-18. Unlike carbon-14, these are not a product of cosmic rays, but a relatively rare component of the Earth's original cargo of oxygen. Heavy oxygen makes molecules of water that are sluggish in their behaviour compared with those containing ordinary oxygen-16, and more noticeably in cool conditions. As a result, the amount of heavy oxygen reaching the ice sheets varies with the temperature.

In the case of Toba, a brief dip in temperature about 74,500 years ago is apparent in the heavy oxygen count in ice of that age from Greenland. But the much bigger plunge towards extreme cold began a thousand years later, and it coincides with one of the most emphatic warmings recorded in Antarctica. For Svensmark such a north–south contrast is, by the way, a sign of cloud-driven climate change. Evidently the effect of intense cosmic rays was much stronger and longer-lasting than any influence of the Toba volcano.

Even if our species did not come close to extinction at that time, abrupt changes of climate due to sudden changes in the Sun's mood repeatedly plagued our ancestors. The bursts of warmth or cold could take effect during one human lifetime. They acted like a long series of intelligence tests, favouring the survival of clever and adaptable people through the opportunities of the warm periods and the hazards of the cold. Archaeologists have still to trace the many links between genetics, migrations, technologies and climate change. But among the thousands of human generations, ours may be the first that was ever frightened by a warming.

Gerard Bond, who refined the studies of the Atlantic ice-rafting events, died in 2005, but the data he left as a legacy still provide the clearest evidence that Nature was capable of working frequent and drastic changes of climate long before the Industrial Revolution. Combined with Jürg Beer's beryllium-10 data, they leave no room for any reasonable person to deny the Sun's important role in climate change, including much of the rise in temperature from the Little Ice Age to the start of the 21st century.

But Beer himself remained opposed to Svensmark's idea that the cosmic rays were more than a symptom of the Sun's mood, and that they exerted a direct climatic effect by their role in cloud formation. This was not a casual preference for variations in solar brightness as the driver of climate. Beer found good evidence of the cosmic rays failing to cause any change of climate when their influx responded not to the Sun's activity, but to changes in another shield that helps to keep the cosmic rays at bay – the Earth's own magnetic field.

'Our results clearly contradict the cloud hypothesis'

Although Edmond Halley is most famous for correctly predicting the return of his eponymous comet, he was a versatile researcher. As the first person to chart the Earth's magnetic field systematically, on oceanographic expeditions, he also grasped that our planet's magnetic poles are always shifting their positions. Before Halley's birth in 1656, compasses on ships in the English Channel pointed east of true north. By 1700 they pointed to the west, and competent skippers steering the traditional course down Channel were liable to wreck themselves on notorious rocks called the Casquets, as in caskets or coffins. Halley urged them to correct the course by a full compass point, or 11.25 degrees.

Fast-forward 300 years and Halley's successors in the science of geomagnetism are worried that the Earth's magnetic field is becoming weaker rather rapidly, after declining gradually for 2,000 years. A French–Danish team compared measurements made by the Danish satellite Ørsted in 2000 with those from the American satellite Magsat twenty years earlier. They found a rate of decline so rapid that the field could by the simplest arithmetic disappear completely in a thousand years or so.

The weak spot over the South Atlantic, where satellites are especially vulnerable to energetic particles coming from the Sun, is growing wider towards the southern Indian Ocean. The decline is bad news for spacecraft engineers. And the Canadian Geological Survey reports that the position of the North Magnetic Pole is shifting at a recently accelerated rate of 40 kilometres a year. Experts wonder if our planet is now getting ready to swap its north and south magnetic poles around.

It has done that very often in the geological past, at

irregular intervals, as recorded in stripes of magnetised rock on the ocean floor and in ancient lava flows on land. The last time your compass would have pointed firmly to the south instead of the north was 780,000 years ago, before the event known as the Matuyama-Brunhes Reversal. On such an occasion the switchover is not quick, because the field weakens for a thousand years or more and then takes a similar time to re-create itself the other way around.

Would it matter if the Earth were to lose, for a while, much of its own magnetic shield against cosmic rays coming from the Galaxy? The answer at first sight seems to be no – no noticeable harm. Magnetic reversals were discovered early in the 20th century by Bernhard Brunhes in France and Motonori Matuyama in Japan. Since then many scientists have looked for evidence of any dramatic effects that might have accompanied the events, but without success. While you can imagine that migrating birds and other creatures that navigate with inborn magnetic compasses must have been confused, little hint of any major climatic change driven by these magnetic switches shows up in the geological record.

Much the same might be said for a weakening of the magnetic field just before the Bronze Age, around 5000 BC. The production of radioactive atoms by incoming cosmic rays was at a higher rate than in any recent century, but again there was no obvious climatic effect. In fact the world was in its warmest phase since the end of the most recent ice age.

Sometimes the Earth seems to try to flip over its magnetic field, but fails. The poles make rapid, widening movements, the field grows much weaker, but then the poles return to their previous ends of the Earth. An event of that kind, detected in volcanic rocks in Chaîne des Puys, France, was the Laschamp Excursion of 40,000 years ago, in the middle of the most recent ice age. The magnetic field waned to one-tenth of its present strength.

A team at the Swiss Federal Institute of Environmental Science and Technology led by Jürg Beer made a special study of this event in ice drilled from the summit of the Greenland ice sheet. They saw the counts of beryllium-10 and chlorine-36 atoms produced by cosmic rays increasing by more than 50 per cent as the field weakened. Despite this increase, no cooling ensued.

What made the result beautiful and compelling was that the indicators of the prevailing climate – heavy oxygen and methane abundances – came from the same layers of ice as the indicators of the flux of cosmic rays. With no need to match the results to data obtained elsewhere, the Laschamp Excursion recorded in the Greenland ice put the supposed link between cosmic rays and climate to a severe test. When Beer and his colleagues reported their result in 2001, they pointed out that, according to Svensmark, the large increase in cosmic rays should have resulted in an increase in global cloud cover and a lower global mean temperature, and this was clearly not the case. Beer was still pressing the point in 2005.

> If the Danish hypothesis is true, the cloud cover in this period should have increased, resulting in a significant climate cooling. … Our results clearly contradict the cloud hypothesis. Since all the parameters are measured in the same ice core, this important result does not depend on how well the ice core is dated.

Here was a powerful argument against any strong connection between cosmic rays, clouds and the climate. Other researchers who supported the idea that the Sun plays an important role in climate change often shared Beer's point of view. Your author Calder was just one of those who, though impressed by Svensmark's thesis, fretted about the problem and spent much time fruitlessly searching for any clear sign of climate changes associated with failures of the Earth's magnetic shield. As a challenge striking at the very heart of the propositions reported in this book, it calls for a clear response. And that means looking in more detail at how the cosmic rays get here.

2 Adventures of the cosmic rays

Remnants of supernovae spray cosmic rays around · They have unexpected roles in the Milky Way Galaxy · Magnetic fields of Sun and Earth repel some of them · The air blocks all but a few energetic atomic particles · Climate-altering cosmic rays scorn the Earth's field

Daring flights in a hydrogen balloon in 1911 and 1912 revealed that the air becomes electrically more conductive the higher up you go. The adventurous Victor Hess of the Institute of Radium Research of the Viennese Academy of Sciences attributed this effect to what he called high-altitude radiation, *die Höchenstrahlung*. Robert Millikan in Chicago re-named the culprits cosmic rays in the mistaken belief that they were gamma rays – the super-X-rays familiar in radioactivity. They soon turned out to be charged particles, including several kinds never seen before.

For four decades, cosmic rays were at the cutting edge of fundamental physics, bestowing Nobel Prizes galore.

When particle accelerators became a less chancy means of discovering new sub-atomic particles, the baton of cosmic-ray research passed to space scientists, who could intercept them in pristine form beyond the atmosphere. And astronomers began to consider their origins, and their role in cosmic housekeeping. At long last, the adventures of the cosmic rays can be related with some confidence, from their fiery origins until their products pass through the air and through our bodies, and disappear into the rocks of the Earth.

Tracking down the breeding grounds

On a wide African plain near Windhoek in Namibia, a group of four telescopes of an unusual kind was nearing completion in 2003. Even before the last of them was ready, telescopes operating in pairs helped to confirm that cosmic rays come from factories in the remnants of exploded stars. For many years scientists had assumed that to be so, but until the new results from Africa the evidence was scanty.

The trouble is that cosmic rays are charged particles. Magnetic fields in the Galaxy and in the vicinity of the Sun and the Earth make them swerve. By the time the particles are detectable in our vicinity, they come almost equally from every direction. Their directions of travel tell you no more about where they started from than does the track of a bluebottle fly.

Astronomers did not despair of tracking down the breeding grounds of the cosmic rays. When they collide with atoms in cosmic space, the products include gamma rays. These should be quite intense where cosmic rays are concentrated in their factories. And as gamma rays are a form of light, they will come from the source to the Earth in just as straight a line as any accompanying visible light.

Compared with ordinary gamma rays emanating from the radioactive elements that litter space and can be seen by satellites, the gamma rays from the cosmic-ray factories have to be about a thousand times more powerful. Their detection calls for large telescopes that capture glows of light in the sky. When the gamma rays hit the atmosphere they create charged particles travelling faster than the speed of light in air, which then produce shock waves of light.

Working on that principle, the Whipple Telescope in Arizona first identified high-energy gamma rays coming from a supernova remnant in 1989. It was the well-known Crab Nebula in Taurus, where a star blew up in AD 1054. But the telescope could not pinpoint the directions from which the gamma rays came – at any rate, not sharply enough to relate them to any particular parts of the Crab's cloud of expanding debris.

Encouraged, but vowing to do better, teams of scientists set about improving their instruments. One is the four-mirror telescope in Namibia, called HESS in honour of the discoverer of cosmic rays. The project brings together scientists from Germany, France, Britain, the Czech Republic, Ireland, Armenia, South Africa and Namibia.

When some of the mirrors were ready, they spent about ten hours staring at a supernova remnant in the constellation of Scorpius, thought to be about the same age as the Crab. As with many objects nowadays, it has a name like a car's number plate, RXJ1713.7-3946, which mainly decodes into its position in the sky. The outcome was the first image of an astronomical object ever seen by very energetic gamma rays.

The picture matched very well with X-ray images showing the shape and size of the expanding shell of debris. In particular, the gamma rays are most intense from one side of the shell, where it has collided with a relatively dense cloud of interstellar gas. That is just where theorists would expect the greatest production of cosmic rays.

The remnant is not small. Even though it lies about 3,000 light-years away, as seen from the Earth it looks wider than the Moon. Paula Chadwick of the University of Durham expressed the team's jubilation about this early result from HESS.

> This picture really is a big step forward for gamma-ray astronomy and the supernova remnant is a fascinating object. If you had gamma-ray eyes and were in the Southern Hemisphere, you could see a large, brightly glowing ring in the sky every night.

If you could also see the cosmic rays, instead of just imagining them, they would be shooting out of the glowing ring in all directions, and then weaving their way through the Galaxy in obedience to its magnetic fields. But as it's only a thousand years old, RXJ1713.7-3946 has barely begun its career as a source of cosmic rays.

Out of the ashes

Although a *nova* means literally a new star, it's really a pre-existing star that suddenly becomes much brighter and therefore more noticeable to sky-watchers. In a *supernova* the brightening is extreme and it signals the demolition of a star in a cataclysmic event. Among various kinds of super-nova, the chief producers of cosmic rays are those of Types II and Ib, in which a star far more massive than the Sun self-destructs.

Deep inside the Sun, a nuclear furnace generates the energy that nourishes life on the Earth, by fusing hydrogen into helium. When most of the hydrogen in the core is exhausted, the helium will burn, making carbon and oxygen. That is as far as a star the size of the Sun will able to go. After shedding its outer layers in a beautiful shroud called a planetary nebula, the core will become a white dwarf star – small, dead and slowly cooling.

In massive stars the nuclear burning by the fusion of elements goes further. Compression of the core by intense gravity raises the temperature high enough for carbon and oxygen to burn, producing still heavier elements. Eventually silicon fuses to make iron, and energy production by nuclear burning reaches its limit. No longer generating heat, and without the power to resist the pressure of gravity, the iron core collapses and brings the rest of the star crashing down on it.

With a huge amount of energy suddenly let loose, the upper layers of the star rebound outwards. Armies of ghostly particles called neutrinos blast most of the star's material into the surrounding space. Meanwhile, nuclear reactions powered by the release of energy create the chemical elements heavier than iron – all the way up to gold and uranium and even beyond.

For a few weeks the supernova blazes as brightly as a billion suns. The dead core that remains is not a white dwarf but a far denser object, a neutron star. The sky is dotted with neutron stars, each marking the death of a massive predecessor. When young they often announce their presence by sending out pulsating radio signals, so they are called pulsars. The best-known supernova remnant, the Crab Nebula, still retains its pulsar in the midst of the star's debris. In many other cases the pulsar gets a sideways nudge, so that it slips away from the wreckage like an arsonist leaving the scene of the crime.

The atomised matter blasted from the stellar explosion

expands freely into space at about one-thirtieth of the speed of light, or 10,000 kilometres per second. As a result it possesses enormous energy of motion, about one-fifth of which will eventually be converted into cosmic rays travelling at velocities close to light speed. But it's not a quick process.

Production of cosmic rays starts in earnest only when the dispersed atomic matter becomes as thin as the gas in interstellar space, and meets resistance from it. The material from the exploded star slows down and mixes with the interstellar atoms. Shock waves become stronger and magnetic fields associated with them are more intense.

Here are the cosmic-ray factories, within the scattered debris. German and Swiss astronomers Walter Baade and Fritz Zwicky first suggested supernovae as the source of cosmic rays in 1934. Fifteen years later, the Italian-born physicist Enrico Fermi at the University of Chicago pointed out that a charged particle in cosmic space could gain energy if it bounced off a moving magnetic field. Think of a slow rubber ball, tossed carelessly by a child, rebounding at high speed from the windscreen of a passing car.

Other theorists soon saw that a shock wave in a supernova remnant makes an especially good accelerator, because irregular magnetic fields in front of the shock and behind it act like mirrors. The charged particles that will become cosmic rays bounce back and forth repeatedly through a shock wave, gaining more energy every time. The magnetic mirrors bottle the particles up, while their acceleration continues. By the time they finally escape from the supernova remnant, the cosmic rays match in their individual energies the products of the particle accelerators on the Earth. Indeed, some are a hundred times more energetic than particles in the very latest machines, but they are relatively scarce.

As hydrogen is the commonest element in the Universe, most cosmic rays are protons – nuclei of hydrogen atoms. Other elements are represented too – helium, carbon, oxygen and so on – roughly in proportion to their abundance in the Galaxy, although an excess of iron is a sign of their origin in supernovae. Such nuances notwithstanding, cosmic rays are just ordinary stuff in very high-speed motion. The most sluggish protons among them travel at about 90 per cent of the speed of light. Their faster colleagues can approach but never quite reach that speed limit. Instead, their energy of motion manifests itself in additional masses.

At Vienna's Institute of Astronomy, Ernst Dorfi has worked out how the timing of events in a supernova remnant depends on the violence of the explosion and the density of gas in its surroundings. In a typical case, his calculations show the expansion starting to slow down about 200 years after the explosion. Half the energy of motion goes into heating the gas in the supernova remnant within 2,000 years. By that time, significant production of cosmic rays has begun, yet it does not reach its peak for 100,000 years, and it continues for hundreds of thousands of years thereafter.

After about a million years the supernova remnant, with most of its energy dissipated, is losing its identity, with only a wandering neutron star to commemorate the once-brilliant blue heavyweight. Meanwhile many other stars will have blown up. At any one time, thousands of supernova remnants are busy handing out their gifts of chemical elements and spraying the Milky Way with galactic cosmic rays.

The tag *galactic*, by the way, distinguishes them from other high-speed particles that you may hear about. *Ultra-high-energy cosmic rays* are rare and probably originate in other galaxies. *Solar cosmic rays* are relatively feeble and come from explosions on the Sun. Often called *solar protons*, they are hazardous for astronauts and spacecraft but seldom have any effect at ground level. *Anomalous cosmic rays* are also feeble. They come from distant shock waves in the Sun's magnetic field and are of interest only to space scientists.

Whenever cosmic rays figure in our story, we mean the ordinary galactic sort. They arrive from outer space as *primary cosmic rays* and generate knock-on products in the air, called *secondary cosmic rays*. These are the ones – the secondaries made by the primary cosmic rays from the exploded stars – that are passing through your skull about twice a second, even as you read this paragraph.

Not just an optional extra

For a long time, most astronomers regarded cosmic rays as curious but unimportant by-products of the death of stars, like litter found after a funeral. By the end of the 20th century a very different perspective was emerging, and in 2001 Katia Ferrière of the Observatoire Midi-Pyrénées, headquartered in Toulouse, wrote a manifesto about it. Her opening sentences promoted cosmic rays to their rightful place in the scheme of things astronomical.

> The stars of our galaxy – traditionally referred to as 'the Galaxy' with a capital G to distinguish it from the countless other galaxies – are embedded in an extremely tenuous medium, the so-called 'interstellar medium' (ISM), which contains ordinary matter, relativistic charged particles known as cosmic rays, and magnetic fields. These three basic constituents have comparable pressures and are intimately coupled together by electromagnetic forces.

When they have left their sources in the supernova remnants, you might expect that cosmic rays travelling close to light speed would soon quit our Galaxy, the Milky Way, and fly off into the wider Universe. The most energetic tend to do so quite quickly, but many of the cosmic-ray particles travel hither and yon within the Galaxy for millions of years, like fishes swimming in a wide but very shallow lake.

The disc of bright stars, which you see edge-on as the Milky Way itself, is squeezed from both sides by gravity. The lines of force of a flattened, sprawling magnetic field thread their way through the disc. Compared, say, with the Earth's magnetism it's very weak, but it operates over many thousands of light-years and it compels the wandering cosmic-ray particles to follow the field lines within the disc. The strength of the field and the numbers of accompanying cosmic rays vary from place to place in the Galaxy. The Sun and the Earth are perpetually on the move, so the local count of cosmic rays changes.

If they try to leave, the field nearly always steers the cosmic rays back into the Galaxy. Just how they eventually leak into intergalactic space is uncertain. For our planet's inhabitants, it's lucky that they do, otherwise the stockpiled cosmic rays might be more than life could tolerate. The average age of the cosmic rays turns out to be 10 to 20 million years. The stockpile has renewed itself hundreds of times since the origin of our planet, but the numbers of cosmic rays in the Galaxy have not remained constant during that time. The birth-rate of explosive stars producing them has varied, and stellar baby booms can be linked to extreme climatic events during the Earth's long history.

The cosmic rays have always hung around for long enough to be an active ingredient in the witch's brew of the Galaxy from which new stars and planets continually form.

By their sheer numbers and momentum the cosmic rays exert a pressure on the gas that occupies the space between the stars. And they help the Galaxy's magnetic field to resist the force of gravity, which tries to drive the interstellar gas towards the midline of the disc and would iron it as flat as the Rings of Saturn, if it could.

The interstellar gas, the magnetic field and the cosmic rays all interact in a slithery manner that leaves them vulnerable to gravity, which can locally reshape the magnetic field and re-route the cosmic rays. As a result, gravity manages to bottle up about half the interstellar gas in relatively dense clouds. So far from being ineffectual, the resistance from the cosmic rays and magnetism ensures that the concentrations of gas will be small and dense enough for eventual star-making.

Dark islands in the Milky Way blot out the view of the stars beyond. These are dust-clouds where the interstellar gas has accumulated stony, icy and tarry grains. Such clouds become maternity wards for new stars and their attendant planets. But there's a lot of chemistry to be done first, and cosmic rays again play a vital role.

In the open, transparent parts of the Galaxy, ultraviolet rays from the stars drive chemical reactions. Elements trickling out from dying stars or blasted from supernovae add extra raw materials to the primordial hydrogen and helium of the cosmos. They combine to make many materials, ranging from water to football-shaped molecules of carbon called buckyballs. But the ultraviolet rays also tend to destroy many of the molecules as fast as they're made.

Only inside the dust clouds, where the veils of stone, ice and tar protect the products from the ultraviolet rays, does the chemistry become fully creative and its products durable. And here the cosmic rays take over from the ultraviolet, as the master chemists of the clouds. They strip electrons from hydrogen molecules and helium atoms, which then become busy – if that adjective is permissible for processes taking tens of thousands of years. Activated hydrogen, for example, interacts with carbon and oxygen atoms to manufacture one of the top players in cosmic chemistry, carbon monoxide.

In these and other ways too intricate to pursue here, cosmic rays share the credit for the creation of the Sun and the Earth, and for our planet's fertilisation by water and carbon compounds from interstellar space. So far from being a trivial by-product or an optional extra in the life cycle of stars, cosmic rays are essential participants in the events.

From Victor Hess's discovery to Katia Ferrière's manifesto, astronomers needed 90 years to come very gradually to this appreciation, and to recognise that cosmic rays even help to mould the galaxies. So perhaps one should be patient with those Earth scientists who still imagine that the third planet of an undistinguished star is far too grand to be influenced in any significant way by piffling little particles from outer space.

How the mother star defends us

The swarming cosmic rays hit the outskirts of the Solar System with a combined punch that has about twice the power of all the starlight seen from the Earth. But once again we're lucky. Clinging like infants to the skirts of the Sun, the planets find shelter inside a huge magnetic field that bats roughly half of the cosmic rays away, back towards the stars.

An understanding of how the mother star defends us followed from the discovery and exploration of the solar wind. It consists of non-stop streams of charged particles

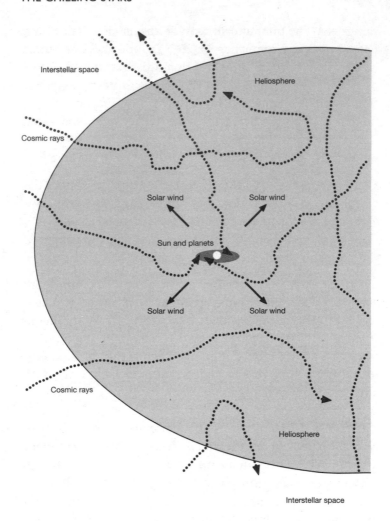

4. *The empire of the Sun extends far beyond the planets in a huge bubble called the heliosphere, blown by the non-stop solar wind. Its irregular magnetic field repels many of the cosmic rays coming from the Galaxy. When this solar shield weakens, more cosmic rays reach the Earth.*

flung from the Sun, and it provides a material link to the Earth's own environment in space. Any idea that the Sun is just a distant ball of light in the sky is completely out of date. We live deep inside its extended atmosphere, which is laced with its magnetic field. A young physicist in Chicago, Eugene Parker, predicted the solar wind with remarkable clarity and detail in 1958. The top experts, as he recalled, greeted the idea with scorn.

> They said to me, 'Parker, if you knew anything about the subject, you could not possibly be suggesting this. We have known for decades that interplanetary space is a hard vacuum, pierced only intermittently by beams of energetic particles emitted by the Sun.'

Yet within four years spacecraft had fully confirmed the existence of the solar wind, and it matched Parker's heretical specifications. Since the 1960s it has been a constant theme in space research, culminating in the joint European–US Ulysses mission, launched in 1990, which has twice flown on a large orbit over the poles of the Sun – a feat never attempted before. Ulysses changed some beliefs about the solar wind, while firming up others.

Armies of protons predominate because the Sun consists mainly of hydrogen. Positively charged atoms of many other elements also show up in the solar wind, together with exactly enough negatively charged electrons to keep the gas electrically neutral. The solar wind drags the Sun's magnetic field with it, so that interplanetary space is filled with magnetism on the move, ready to fight the cosmic rays.

The wind blows at roughly 350 or 750 kilometres per second, depending on what region of the Sun it emerges from. Even the fast windstream is much slower than the cosmic rays. Particles in the solar wind cross the Earth's orbit a few days after leaving the Sun's atmosphere. Then they continue for a year or two in their flight away from the Sun, blowing a huge bubble into interstellar space, called the heliosphere.

Eventually the solar wind, spreading far and wide, becomes so diffuse that the interstellar gas can successfully resist it. It then comes to a halt at roughly five times the distance of Neptune, the most remote of the major planets. That boundary of the Sun's empire is so far out that light, or any unimpeded cosmic-ray particle, takes about twenty hours to come in from it, compared with eight minutes for light to reach the Earth from the Sun.

The size of the heliosphere depends on how forcefully or otherwise the solar wind has blown in the previous couple of years. A scarcity of dark sunspots blemishing the bright face of the Sun is a sign that it's in a relatively quiet state. At such times the density of the solar wind decreases, but because its average speed increases, the impact pressure pushes the outer boundary of the heliosphere a little further away.

Over timescales of millions of years the Solar System encounters clouds of interstellar gas a hundred times denser than in our present location among the stars. The increased pressure then squeezes the heliosphere until it fails to inflate even as far as the outer planets. On the other hand, when the Sun was in its infancy, the solar wind was far stronger than today and the heliosphere extended much further.

The Sun rotates on its axis every four weeks, and as a result the magnetism pulled from the Sun by the solar wind fills the heliosphere with mobile lines of force in spiral form. Near the Earth they slant in from the west at 30 to 45 degrees from the direction of the visible Sun. The main work of deflecting and in many cases repelling the cosmic rays is done, not by these regular field lines, but by intense small-scale irregularities in the magnetic field that scatter the cosmic rays and drag them along with the wind, to some extent.

Collisions between fast and slow windstreams coming from different parts of the Sun are one cause of shocks that create these irregularities. Others are due to magnetic explosions in the Sun's atmosphere that throw out huge puffs of gas, called mass ejections, which appear as strong and sudden gusts in the solar wind. During years of solar storminess, indicated by regions of intense magnetic activity that make large numbers of sunspots, the shocks are more numerous and powerful.

The *eminence grise* of the Ulysses solar-polar mission, John Simpson of Chicago, had a favourite analogy for the experience of cosmic rays trying to enter the inner Solar System.

> Imagine you're rolling tennis balls down an upward-moving escalator. Some of them will bounce back off the steps. Now speed up the escalator to simulate increased solar activity and many more of the tennis balls will come back to you. Fewer will reach the bottom of the escalator.

And only about half the number of cosmic rays will reach the inner Solar System, where the Earth circles around the Sun. Many of those lost when the Sun is most active are particles of relatively low energy that are in any case repelled by the Earth's own magnetism. Nevertheless, over the past half-century a ground station set up by Simpson at Climax, Colorado, showed counts of cosmic rays of moderate energy, averaged month by month, being reduced by 25–30 per cent following periods of high sunspot counts. There is usually a delay of a year or two from maximum sunspots to minimum cosmic rays, because of the time taken for the shock waves to travel out through the heliosphere.

The last lines of defence

After a day or two spent zigzagging through all the magnetic shocks put in their way by the solar wind, some of the cosmic rays eventually approach the Earth. The penultimate obstacle in their way is our home-grown magnetic shield. A dynamo in the liquid iron core of the planet makes most of the Earth's magnetism, which occupies a bubble within the solar wind called the magnetosphere. Gusts in the solar wind deform the magnetosphere and cause magnetic storms, when compass needles wander and auroras glow in the polar skies.

Beginning in 1868, scientists in Greenwich and Melbourne monitored the variations in magnetism on opposite sides of the Earth. The recording of what is called the *aa* index continues to this day, now using geomagnetic observatories at Hartland in England and Canberra in Australia. Unwittingly, the pioneers were measuring the vigour of the solar wind, to provide a wonderful record of the connection between Earth and Sun going right back to Victorian times. But there has been a lot of confusion about it, leading to exaggerations of the role of the Earth's magnetism in protecting the Earth from cosmic rays.

When a large mass ejection from the Sun passes by the Earth, it acts like a magnetic umbrella. The cosmic-ray count can plummet by perhaps 20 per cent in less than a day and then take several weeks to return to normal. The discoverer of such decreases, Scott Forbush of the Carnegie

Institution in Washington DC, thought that storms in the Earth's magnetic field were responsible, when in fact these were just simultaneous by-products of the same solar events. Spacecraft visiting other parts of the Solar System report Forbush decreases in cosmic rays from time to time, so confirming that they are the result of individual shock waves in the heliosphere.

When they encounter the Earth's own magnetic field, individual particles still follow erratic tracks, as if they were in a pinball machine. The filtering effect is nevertheless quite simple in its outcome. The energetic particles can reach the Earth anywhere, although they need higher energy over South-East Asia than over Brazil and the South Atlantic, because the magnetic field is lopsided. Cosmic rays of somewhat lesser energy are excluded altogether from the equatorial zone, because that is where the magnetic field is parallel to the Earth's surface. The least energetic particles are either denied access altogether or else forced to come down nearer to the magnetic poles, where field lines slant steeply and lead the cosmic rays inwards.

The primary cosmic rays arriving from starry space come to a sudden sticky end when they hit the Earth's atmosphere. They might as well crash into a massive castle wall. To ground-dwellers, the air 25 kilometres up would seem extremely thin, yet already it's far denser than anything the particles have ever encountered during their multi-million-year wanderings through the Galaxy – otherwise they wouldn't have survived this far.

The planet Mars is an object lesson in the air's importance as our last line of defence against the cosmic rays. The much skimpier Martian atmosphere offers no more protection from cosmic rays than does the Earth's air above 20 kilometres. As a result, astronauts walking on the surface of

Mars will be exposed to as much potentially harmful cosmic radiation in just one day as most inhabitants of our planet receive in a year. Space agencies are accustomed to hardening the electronics of spacecraft against damage by cosmic rays, and a NASA team reporting on health risks on Mars in 2005 was a little rueful on the subject.

> The most relevant examples of this success are the Mars Exploration Rover vehicles, which have operated on the surface of Mars for over a year … Radiation effects on electronics are at present much better understood, and more preventable, than effects on living organisms.

Encounters with the Earth's air bring to a halt virtually all of the incoming primaries – the high-speed protons and nuclei of heavier atoms – long before they can reach the ground. The swarms of secondary cosmic rays that replace them are swiftly moving particles released in atomic and nuclear interactions. These go on to have further encounters of their own, with the result that energetic primaries can initiate showers of millions or even billions of particles. Physicists have fun tracing complicated sequences of events that involve various kinds of sub-atomic particles plus gamma rays, but very few of the products reach the lower levels of the atmosphere.

Judging by their ability to knock electrons out of molecules in a test instrument, the intensity of the cosmic rays increases after the initial impacts, because of the large number of secondary particles produced. It reaches a peak about 15 kilometres above the ground, where the intensity is about twice that of the primary cosmic rays just before they hit the atmosphere. The air's blockade is then so effective that at sea-level the intensity is reduced to about one-twentieth of the peak. Starting from the ground, Victor

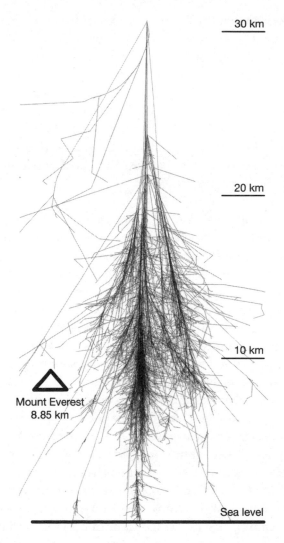

30 km

20 km

10 km

Mount Everest
8.85 km

Sea level

5. When an energetic cosmic-ray particle hits the Earth's atmosphere it produces a shower of sub-atomic particles of many different kinds. Nearly all of them are stopped by our aerial shield and only a few reach the lowest altitudes. (CORSIKA calculations by Fabian Schmidt, University of Leeds)

53

Hess realised that the cosmic rays must be coming from the sky because their intensity kept increasing, the higher he went in his hot-air balloon.

Female aircrew are often grounded if they are pregnant, to protect their babies from possible interstellar harm. Passenger aircraft cruise routinely 10 kilometres above sea-level, which is twice the altitude reached by Hess in his balloon. Those flying by trans-polar routes are especially prone to the cosmic rays channelled down the slanting lines of the Earth's magnetic field.

People living at high altitude also experience increased radiation from the sky. At more than 3,600 metres, the world's highest capital city is La Paz in Bolivia, and there the cosmic-ray intensity is about twelve times greater than in nearby Lima in Peru, which is only 150 metres above sea-level. Some 8 million people live on the altiplano, among the high Andes, and the Incas and their predecessors thrived there for millennia before any Europeans arrived. So the high radiation levels cannot be too deadly.

Averaged world-wide, the medically significant cosmic radiation is scarcely greater than that from the radioactivity in food and water, and it amounts to only 16 per cent of all naturally occurring atomic radiation to which human beings are exposed. Along with natural radioactivity and the effects of bodily heat and chemistry, cosmic rays contribute to genetic mutations that can cause birth abnormalities and cancers, but they also make possible the evolution of species. As related later in this book, the indirect effects of cosmic rays via climate change may be more important in evolution.

'Who ordered that?'

The sub-atomic mayhem in the upper air produces just one kind of charged particle that reaches the Earth's surface in large numbers and with only a moderate loss of energy. It's called the muon and, surprisingly, physicists didn't notice it until 1937. Then Isadore Rabi of Columbia University, New York, expressed their bewilderment. 'Who ordered that?' he asked.

No atomic theorist up till then had guessed that the electron, the lightest of charged particles, might have a big brother. But that's what the muon is – similar to the electron in every respect except for its mass, being 200 times heavier, and for its instability. Created during the decay of a pion, a nuclear force particle that is mass-produced in the early impacts on the air, the muon itself survives for just 2 millionths of a second until it sheds two ghostly neutrinos and becomes an ordinary electron.

If you had to invent a sub-atomic particle to sneak a message from the stars through the barrier of the Earth's atmosphere, you'd not improve on the muon. Ordinary electrons won't do. Although they are present in both the primary and secondary cosmic rays, they are too lightweight – too easy to impede and then to seduce into ordinary chemical interactions with the molecules of the air. On the other hand, the much heavier protons and their neutral siblings, the neutrons, interact too readily with atomic nuclei in the molecules, shedding energy rapidly and engaging in nuclear reactions. For every 1,500 protons and neutrons available in the cosmic rays 15 kilometres up, only one reaches sea-level.

For an infiltrator you need a particle with little inclination to react with anything, light enough to be mass-produced from the energy available, and yet having sufficient momentum to carry it through all those air molecules snatching at it like harpies. The muon meets all the specifications. In doing so, it joins the carbon atom, hydrogen bonding and the water molecule in a select group of products of the physical Universe that are extraordinarily well crafted to play key roles on a living planet.

To take effect the muon needs a helping hand from Albert Einstein. It's so short-lived that it would travel for only 600 metres through the atmosphere, with no hope of reaching the ground, were it not for the relativistic trick that stretches time for high-speed travellers. At almost the speed of light, the muon's internal clock runs so slowly that its nominal life of 2 millionths of a second can be stretched a hundredfold or more. Thanks to this dispensation, the muons take in their stride the journey down to sea-level, where they make up 98 per cent of the secondary cosmic rays. Most of the rest are a few survivors from among the protons and neutrons.

The muons continue on their way into the water and rocks of the Earth. When physicists are searching for more elusive sub-atomic particles, they need to escape from the cosmic rays. So they put their experiments into deep mines or tunnels. Even there, a few of the most persistent muons still show up as noise in the instruments.

For Svensmark, the muons are the cosmic rays that most affect the climate. That's precisely because they reach to the lowest levels of the atmosphere where they can influence the formation of low-level clouds that cool the world. So when he wanted to respond to the claim that cosmic rays could not be responsible for climate change, it was on the origin of the muons that he focused his attention.

Confirming a hunch

The argument of Jürg Beer, with which the previous chapter ended, was that some large variations in the influx of cosmic rays, signalled by production rates of atoms of carbon-

14 and beryllium-10, were not accompanied by noticeable changes in the climate. The Laschamp Event 40,000 years ago, when the Earth's magnetic field almost disappeared and the count rates shot up, was Beer's prime example.

The conundrum heartened Svensmark's critics. While it remained for several years as a contradiction at the core of the story about cosmic rays and climate, Svensmark himself had a hunch about what the solution might be.

He suspected that cosmic rays that leave behind the carbon-14, beryllium-10 and other business cards as records of their passing are in some important sense different from the cosmic rays that penetrate to the lowest levels of the atmosphere. That might explain why the reported variations in cosmic rays are not necessarily reflected in changes in the abundance of low clouds – those that are most involved in changes of climate due to cosmic rays.

It was an idea strange enough to sound like special pleading, and Svensmark was too preoccupied with mounting a laboratory experiment to break off to validate his idea. Not until 2006 was he able to take it up in brief intervals at home. Even then he had to recruit his son Jacob, a physics student, as an unpaid assistant.

Their family project took advantage of a renaissance in cosmic-ray research that came with the pursuit of particles of ultra-high energy arriving from the Galaxy or beyond. These produce extensive showers of secondary sub-atomic particles raining through the atmosphere. By 2005, the multinational Pierre Auger Observatory in western Argentina was celebrating early results from an array of detectors spread over 3,000 square kilometres.

Smaller observatories for extensive air showers in other parts of the world include a German array of 252 instrumented stations called KASCADE, which is short for the

Karlsruhe Shower Core and Array Detector. In 1989 the Forschungszentrum Karlsruhe made available a computer program for tracing the behaviour of cosmic rays in the atmosphere, and Dieter Heck has kept upgrading it ever since. The program has its own acronym, CORSIKA, meaning Cosmic Ray Simulations for KASCADE, and it calculates the complex sub-atomic changes and reactions occurring after the primary cosmic rays hit the Earth's atmosphere. Effects of the increasing density of the air as the particles descend are taken into account, and so is the influence of the Earth's magnetic field.

Knowledge of dozens of different kinds of sub-atomic particles, accumulated by physicists over 100 years, is built into the program. In the behaviour of the particles there is a large element of chance – when they will break up, for example, or whether they will interact with other particles. So CORSIKA tries out many possibilities using random numbers generated by the computer. For obvious reasons, statisticians call this the Monte Carlo method.

Because the aim of CORSIKA is to calculate what particles will eventually arrive at the detectors on the ground, it connected directly with the meteorological issue. Svensmark, too, wanted to know about the relatively few charged particles that survive into the lowest part of the atmosphere. The most important of them are the muons made in interactions high in the atmosphere but travelling fast enough to have their very short lives extended sufficiently by Einstein's time-stretching for them to be able to reach sea-level.

To achieve the necessary lifetimes, the muons have to be manufactured by very energetic cosmic rays hitting the Earth's atmosphere. Such incoming primary particles are relatively scarce, but that is offset by their large effective masses, when relativity can multiply the ordinary mass of an energetic proton (hydrogen nucleus) a hundredfold. That represents energy available for creating great showers of secondary particles, including large numbers of energetic muons. Just how scarcity and productivity might balance out was, for Svensmark, a question that CORSIKA could answer.

It was such a big and unwieldy program that he needed help in installing CORSIKA and running it, and that was where Jacob Svensmark came in. As father and son explored the effects of cosmic rays of different energies, all their spare time in the month of May 2006 went into repeated runs of the program, each taking from hours to days. The results were compelling.

The calculations focused on the cosmic-ray activity in the lowest 2,000 metres of the atmosphere, which contributes to the formation of the climatically important low clouds. Remarkably, as many as 60 per cent of the all-important muons are products of cosmic rays coming in from the Galaxy with so much energy that the Sun's magnetic defences offer no protection against them. They are not involved in the climatic variations over the centuries due to changes in the Sun's behaviour. Over millions of years, the input of these energetic cosmic rays changes as the Earth travels with the Sun through changing scenery in the Galaxy. Later in the book we'll see how big the consequences were for the Earth's climate.

The remaining 40 per cent of cosmic rays that influence events in the lower air are subject to control by the Sun's magnetic field. That's plenty to account for the swings between warm and cold periods already described. But the Earth's magnetic field has a much weaker effect. The outcome of the calculations is that only 3 per cent of the

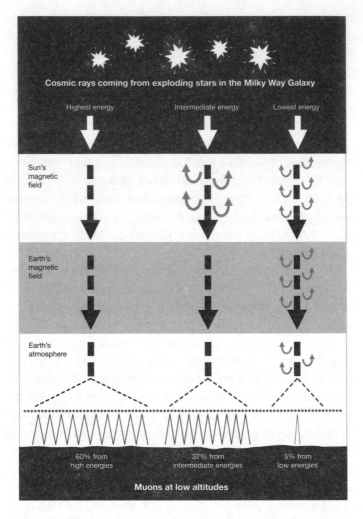

6. *The most important cosmic-ray particles that assist in cloud-making in the lower atmosphere – the muons – originate mainly from particles that arrive from the stars with very high energy. The magnetic defences of the Sun and the Earth have little effect on them. The Sun's magnetic field influences the supply of a large minority of muons, but few are obedient to changes in the Earth's magnetic field.*

cloud-making muons that penetrate to low altitudes originate from incoming particles of such relatively feeble energy
that changes in the Earth's magnetism can affect them. On
the other hand, most of the business cards like beryllium-10
are made at higher altitudes, by incoming cosmic-ray particles of moderate energy that respond to variations in the
Sun's magnetic field. In many cases they are also strongly
influenced by the Earth's magnetic field. Reduce the Earth's
magnetism drastically, as happened in the Laschamp
Event, and the count of beryllium-10 and chlorine-36 atoms
by Beer and his team goes up by more than 50 per cent. Yet
the climate-changing muons would increase by no more
than 3 per cent if the magnetic field disappeared entirely.
Svensmark's hunch about different kinds of cosmic rays is
now confirmed in precise terms.

Laschamp revisited

Although Svensmark believes that these results from
CORSIKA dispose of the main issue against his cloud
hypothesis, as raised by Jürg Beer, the details of what the
cosmic rays and the climate were doing around Laschamp
time may deserve re-investigation. The warming reported
by Beer as coinciding with a sharp rise of his tell-tale atoms
may mean that the Sun's magnetism was strengthening – to
cut the low-altitude cosmic rays and reduce cloudiness – at
the same time as the Earth's magnetism was weakening
and allowing more production of carbon-14, beryllium-10
and so on.

That is not impossible. In the general run of ice-core
records of climate, the warming in question is seen to be
one of the Dansgaard-Oeschger Events that repeatedly
pushed up temperatures dramatically during the last ice
age, and were undoubtedly the result of increased solar

activity. But the busy Sun would also repel many of the cosmic rays of lower energy too. Without that effect, the rise in atomic counts when the Earth's magnetic field faltered at Laschamp time might have been even bigger.

Archaeologists using radiocarbon to date their finds from the middle of the ice age have had to sort out the effect of increases in carbon-14 production when the magnetic field weakened. It gave rise to errors of up to 5,000 years in their dates. In 2004 a team led by Konrad Hughen of the Woods Hole Oceanographic Institution in Massachusetts published refined carbon-14 data from the sea floor off Venezuela.

They saw the Laschamp peak in carbon-14 production occurring relatively briefly around 40,500 years ago, followed by a large and almost non-stop fall to 37,000 years ago. That brings you to the time of the peak warming in Beer's data, which also coincides with a drop in beryllium-10 and chlorine-36 production. Perhaps any remaining mismatch between cosmic rays and climate will dissolve away as the data improve.

With Laschamp, our story gets ahead of itself. Beer's was the most persuasive objection, scientifically speaking, since Svensmark first proposed in 1996 that cosmic rays directly influence the climate. To deal with the challenge early in the book seems prudent, in case a well-informed reader might be deterred from pursuing the theory further. But it's high time to turn to the discoveries about the other main players in our cosmic drama, the clouds.

3 · A shiny Earth is cool

*Fashionable climate science is baffled by ordinary clouds ·
Satellites show cloudiness varying with cosmic-ray counts · Most
affected are the low-level clouds that cool the world · The clouds
confirm their authority by warming Antarctica · The discoveries
make drastic global warming less likely*

To a very large extent the clouds *are* the weather, to casual
onlookers and meteorologists alike. Over our heads and
on the horizon they enact a ceaseless show of sunshine or
gloom, calm or storm, rain or snow, with seldom the same
scene twice. In their most alarming forms they spit lightning, whip up tornadoes, or build the great cloud-walls of
hurricanes.

Anyone wanting to describe, categorise, analyse and
interpret the clouds is soon puzzled by their capricious-
ness. To feign madness, Shakespeare's Hamlet took his cue
from a cloud, saying that it looked like a camel, then a
whale, and finally a weasel. Centuries later, clouds still vex
the forecasters who calculate tomorrow's weather and
long-term climate change in supercomputers.

Networks of computing points 100 kilometres apart cover the world, but individual clouds fall through the gaps like small fry through a fisherman's net. The computer model-makers have to rely instead on theories about the average behaviour of clouds. Their attempts to predict climate change region by region give contradictory results, according to how the models deal with the clouds. In 2004 Kevin Trenberth, a leading modeller at the US National Center for Atmospheric Research, was candid about it.

Climate models do not do clouds well – they are perhaps the biggest problem we have in using climate models to make predictions about global warming.

Just how silly the computer models are became clearer in the following year. Institutes in France, Germany, the UK and the USA compared the performance of ten models of the atmosphere with satellite observations of actual clouds from 1983 to 2002. Several of the models grossly underestimated the cloud amounts at middle and low altitudes. The report by Minghua Zhang of Stony Brook University and his colleagues was an admission of failure.

For individual ... cloud types, differences of seasonal amplitudes among the models and satellite measurements can reach several hundred per cent.

Pleas from the climate scientists for better facts about clouds were answered, with a pair of satellites that went into orbit in April 2006 on a three-year mission. The French–US Calipso and NASA's CloudSat fly in company, looking down on the same clouds within fifteen minutes of each other, one with laser beams and the other by millimetre radar. They can make out the different layers within thick clouds, measure the sizes of droplets, and distinguish those that are falling as rain. All this and more they accomplish for the very first time, and a comment by Graeme Stephens of Colorado State University summed up the depth of ignorance before they flew.

The new information from CloudSat will answer basic questions about how rain and snow are produced by clouds, how rain and snow are distributed worldwide, and how clouds affect the Earth's climate.

These confessions came when huge efforts were going into climate modelling by supercomputers that attempted to predict the climate 100 years ahead. Despite the discrepant results, some scientists proclaimed an impending climate catastrophe due to emissions of carbon dioxide. The human species, they said, should stifle its industries or suffer the consequences. Several scientific academies and leading scientific journals affirmed that the science of climate change was settled. Researchers with a better grasp of the role of the clouds in climate change were often spurned.

Anyone familiar with warm tropical nights has experienced the greenhouse action of vapours and gases in the air. The Earth's surface radiates heat into space in the form of invisible infrared rays. As a result, deserts can become quite chilly after dark. But in the humid tropics, the water molecules overhead intercept the heat and radiate it back towards the surface, so you can sip your nocturnal rum in your shirtsleeves.

That is the natural greenhouse effect. Due mainly to water vapour, it's essential for keeping the planet's surface congenial for life. Carbon dioxide works in a similar way, which is why there has been concern about the rising amounts of that gas in the air. What is at issue is how great the warming effect will be, as carbon dioxide continues to increase. The direst predictions become questionable when you take into account the real role of the clouds.

Don't imagine that diligent work on the cloud calculations, aided by better satellite data and more billions of dollars, will one day bring the computer models into closer agreement with one another and with reality. The flaws go much deeper than any technical details of the programs. In climate changes supposedly driven mainly by the rising levels of carbon dioxide, the modellers' belief is that the clouds are passive participants.

This chapter will explain that the clouds are in charge of the climate. Their variations follow changes in the intensity of the cosmic rays due to stronger or weaker magnetic shielding by the Sun, with little regard for anything else that may be happening on the Earth. The kinds of clouds that matter most in climate change can be identified. And confirmation that the clouds really are in the driving seat comes from strange goings-on around the South Pole.

The budgets of the clouds

In the daylight hours of summer, the cooling effect of clouds is an obvious fact of life. However grey they look from below, once you get above the clouds, on a mountain or in an aircraft, they gleam white. They scatter back into space about half of the local incoming sunshine that would otherwise warm the Earth's surface below. In addition, some of the solar radiation is absorbed within the clouds.

Common experience also tells you that cloudy nights tend to be less chilly than starry nights, which in winter are often frosty. The clouds exert a greenhouse effect of their own, by intercepting heat escaping from the Earth's surface. Although they also radiate infrared rays out into space, the cloud tops are colder than the ground so the loss of heat is less.

The pluses and minuses of the warming and cooling effects of clouds over the planet as a whole make for a quite complicated budget of incoming visible light and outgoing infrared rays. It was largely a matter of conjecture until special-purpose instruments went into space on three American satellites, launched in 1984 and 1986. They measured the incoming sunlight and the outgoing infrared rays, worldwide. By the early 1990s the results of NASA's Earth Radiation Budget Experiment were clear.

Taken altogether, clouds are strong coolers. The exceptions are thin clouds, which have an overall warming effect. The high feathery cirrus clouds are so cold, at around minus 40 degrees Celsius, that they radiate into space much less heat than they block going out from the Earth. The most efficient coolers, on the other hand, are thick clouds at middle altitudes, but they occur over only about 7 per cent of the Earth at any one time.

Covering nearly four times as much of the surface are the low-level clouds. They account for 60 per cent of the total cooling. As well as barring the sunshine, their relatively warm cloud tops radiate heat efficiently into space. And among the low clouds, the most important coolers of all are the wide and flat blankets of stratocumulus that sprawl across 20 per cent of the Earth's surface. They occur mainly over the oceans, where they provide monotonous scenery for passengers on intercontinental flights.

Overall, the clouds of the world cut the warming effect of the incoming sunshine by 8 per cent. If nothing else changed, removing this huge parasol would raise the planet's mean temperature by about 10 degrees Celsius. Conversely, an increase in the low clouds by only a few per cent would chill the world noticeably.

With more clouds, astronauts in space see the Earth

gleaming more brightly. Astronomers on the ground can observe the sheen too, when they look in the mirror of the Moon for the ghostly earthshine that lights the parts shadowed from direct sunlight. The shinier the Earth, the cooler it is, simply because it is throwing away more of the Sun's warming rays.

The average cloud cover changes from year to year. Diligent chroniclers of the weather have recorded local variations for centuries, but a worldwide view of the clouds became possible only with the first weather satellites. By showing the complete drama of the global weather as it unfolded beneath their cameras in space, they revolutionised meteorology. Since 1966 they have provided forecasters with continuous operational services of ever-increasing quality and coverage. Television viewers learned to recognise animated satellite images of rain-clouds or hurricanes heading their way.

From 1983 onwards the International Satellite Cloud Climatology Project pooled the data coming in from the civilian weather satellites of all nations. Masterminded by William Rossow in NASA's Goddard Institute in New York, it produced monthly charts of average cloud cover divided into squares of the Earth's surface about 250 kilometres wide. The charts beautifully depicted the changing seasons, and the monsoons edging in to cover southern Asia with a vast duvet of clouds. During the climatic interludes called El Niño, the satellite compilations logged large changes in cloud distribution over the tropical Pacific and South America. They also revealed a connection between global cloudiness and the rhythms of the Sun.

The missing link between the Sun and the climate

At Christmas 1995 the Danish Meteorological Institute on the northern outskirts of Copenhagen was almost deserted except for the forecasting office. One light burned on another floor, where Svensmark was pursuing an idea about clouds so compelling that he neglected his wife and young sons over the holiday period. Until then he had not heard of William Rossow's compilation of the satellite data on clouds, but when he found it on the internet that Christmas it helped him to reveal a previously unknown way in which the Sun influences the Earth's climate.

Svensmark was due to move to another division in the institute in the New Year. He would join Eigil Friis-Christensen, who was in charge of solar-terrestrial physics and had a long-standing interest in magnetic storms, auroras and their apparent link to variations in the iciness of the sea around Greenland. With Knud Lassen, another old Green-land hand who was his former boss, Friis-Christensen noticed a curious match between the rise in temperatures in the Northern Hemisphere during the 20th century and a quickening of the sunspot cycles.

When they announced this result in 1991, Friis-Christensen found himself cast in the role of the chief spokesman for the role of the Sun in climate change. That had been under consideration for nearly 200 years, since the astronomer William Herschel in England noticed that the price of wheat was higher when sunspots were scarce. But by the 1990s, many climate scientists had concluded that the Sun didn't matter much. Measurements by spacecraft of variations in the intensity of sunlight suggested that they were sufficient for only a small influence on the climate.

Unknown to Friis-Christensen, his new recruit began

using his spare time towards the end of 1995 to follow up a hunch about how another effect of solar variations might be much stronger. Svensmark thought that the supply of cosmic rays that the Sun admits to the Solar System might help to control the Earth's cloudiness. More cosmic rays, more clouds. Scientists in Russia had flirted with the opposite idea, that cosmic rays might reduce cloudiness. Either way, the connection between the stars and the clouds was not easy to pin down.

As soon as Svensmark gathered a few data from the World Wide Web he saw changes in cloudiness from year to year that seemed to follow the variations in cosmic-ray intensities. In mid-December he showed some early results to Friis-Christensen, who assured him that the proposition that cosmic rays could increase cloudiness was a new one and, furthermore, that it was not unreasonable.

In fact it was just the kind of mechanism that Friis-Christensen had spent several years looking for, which could amplify the influence of the Sun in climate changes. When Svensmark changed jobs in January they would research it together. It would no longer be a spare-time hobby, but full-time paid work. Well, more than full time.

The encouragement from his future boss led Svensmark to cancel his Christmas holiday and look for better data on cloudiness. He had been making do with cloud results from US Air Force weather satellites and NASA's general-purpose Nimbus spacecraft. When his surfing led him at last to the International Satellite Cloud Climatology Project, and he could draw on Rossow's very detailed results for the period mid-1983 to the end of 1990, the investigation proceeded quickly.

The use of different types of weather satellites from several countries, and difficulties in distinguishing clouds from icy, cold and mountainous surfaces, created imperfections and uncertainties in the satellite compilations. Svensmark elected to use only the monthly records of clouds over the ocean as seen by American, European and Japanese geostationary satellites hovering high over the Equator. From various possible sources of cosmic-ray data he selected the monthly averages of neutron counts at John Simpson's station at Climax, Colorado.

The match was striking. Between 1984 and 1987 the Sun gradually became less stormy and more cosmic rays reached the Earth. Cloudiness over the oceans increased progressively by nearly 3 per cent. Then the cosmic rays declined till 1990 and the cloudiness decreased too, by 4 per cent. The results suggested that variations in cloud cover due to cosmic rays could have much more effect on the Earth's temperature than the small variations in the intensity of light coming from the Sun.

The clouds obeyed the cosmic rays closely. By the norms of climate science the correlation was exceptionally good, and Svensmark and Friis-Christensen were astonished that no one had noticed such an obvious linkage before. Afraid of being beaten to the announcement of the discovery by other scientists, they rushed to complete a scientific paper. It went off at the end of February 1996 to the journal *Science* in Washington DC.

Instead of the quick publication they were hoping for, queries came back. When those were dealt with by brief additions, the verdict was that the paper had become too long for *Science* and should be published elsewhere. Friis-Christensen then tackled the editor of the *Journal of Atmospheric and Solar-Terrestrial Physics*, hoping for fast-track treatment. The expanded paper was published there, though not until the following year.

Meanwhile the organisers of a gathering of space scientists, to be held that summer in Birmingham, England, had in all innocence invited Friis-Christensen to give a short talk on solar effects on climate. With Svensmark's agreement, he decided that, whatever the response from the second journal, he should summarise the cosmic-ray connection during his talk. So it came about that the first printed account to be made public was a short press release requested by Britain's Royal Astronomical Society, which looked after the media in Birmingham.

The title was 'The missing link in Sun–climate relationship'. Together with Friis-Christensen's talk and related interviews, the press release provoked a short-lived flurry of interest, which was prolonged only in Denmark. More typical was *The Times* of London, which carried a brief report tucked away on an inside page under the headline, 'Exploding stars "may cause global warming"', with quotation marks distancing the newspaper from the story.

At the Birmingham meeting, Calder monitored the reactions with professional concern. He had learned by chance what Svensmark and Friis-Christensen were up to, and with their help he was already writing a book about the Sun and climate change. He feared that when his fellow science writers woke up to the discovery of the cosmic-ray connection with cloudiness, the story might seem old hat by the time his book was out. He needn't have worried. Outside Denmark, Calder had the subject virtually to himself – not just until *The Manic Sun* was published in April 1997, but for years afterwards. It was the news that no one wanted to hear.

As for Svensmark, he knew that he faced a battle to get the discovery accepted – although he did not anticipate that it would develop into a war lasting ten years or more.

Scientifically he had to confront the natural world, and from a tangle of wiggling records of cosmic rays, solar storminess and terrestrial cloudiness he would tease out the fine details to consolidate the story. But the war was on two fronts, because his ideas were generally attacked or disregarded by the scientific community.

'Naïve and irresponsible'

Any scientist with an original idea expects strong criticism, while colleagues and rivals try to prove the data or the theory wrong. It's the way science works, by weeding out errors until only well-substantiated conclusions survive. An idea that is met with a chorus of opposition is usually wrong. On the other hand, there are many histories of a genuine discovery being resisted too hard, or a false consensus persisting too long. The process is uncomfortable because scientists are passionate human beings, not logical robots. Normally the debate proceeds with some decorum, but climate science became bad-tempered.

The new mood was apparent in the Intergovernmental Panel on Climate Change, which in 1990 began to issue warnings of imminent overheating of the planet. Its predictions depended on attributing the modest rise in global temperatures during the 20th century to increasing amounts of carbon dioxide in the air. Any suggestion that natural factors such as solar activity might have been largely responsible was unwelcome.

The Danish delegation to the Intergovernmental Panel on Climate Change made a modest proposal in 1992, that the influence of the Sun on the climate should be added to a list of topics deserving further research. The proposal was rejected out of hand. And when, in 1996, a Danish newspaper invited the panel's chairman Bert Bolín to comment on Svensmark's results on cosmic rays and clouds, as announced by Friis-Christensen at the Birmingham meeting, he was scathing: 'I find the move from this pair scientifically extremely naïve and irresponsible.'

Those were unusual words for a professor of meteorology from Stockholm to use about a report delivered by a professor of physics from Copenhagen. Within the Danish Meteorological Institute itself, Svensmark encountered bad temper at the personal level. Opposition to his ideas was sometimes aggressive, even in the canteen. Some colleagues did not wish to rub shoulders with anyone not wholly captivated by the hypothesis that man-made carbon dioxide was the main driver of climate change.

They arranged to roast Svensmark later that year, when scientists from Nordic countries gathered for a meeting in Elsinore. Strictly by way of entertainment, the organisers invited Svensmark to give a talk about cosmic rays and clouds, after a bibulous conference dinner, so that everyone could scream at him. They did so with gusto.

A substantive question emerged from the ridicule. Wasn't Svensmark crazy to suggest that cosmic rays might affect cloud formation? Prominent in the audience was Markku Kulmala from the University of Helsinki, chairman of the International Commission on Clouds and Precipitation. He listened in silence until someone called on him to explain why Svensmark's idea was wrong. Kulmala's brief remark then took everyone aback: 'It could be right.'

Unsatisfied, the questioner protested that Svensmark's research was 'dangerous'. That was again a curious word to use, to describe theoretical studies that involved no poisons, projectiles or explosions. The only danger that could conceivably arise would concern scientific beliefs and public policies, should Svensmark's ideas show that the assumptions about global warming and its causes were faulty.

Official funding agencies in Denmark were reluctant to support Svensmark's deviant research. Help came instead from the Carlsberg Foundation, which since the 19th century had happily channelled profits from lager into exciting scientific research of many different kinds. It ignored a letter from a senior government scientist to the foundation's director, which urged him to cancel the grant. Even when Svensmark won Danish prizes for his discovery – the Knud Højgaard Anniversary Research Prize and the Energy-E2 Research Prize – sections of the press were scandalised.

Thanks to the Carlsberg funding, a new pair of eyes joined Svensmark's in the hunt. Nigel Marsh, who came from Britain, had recently gained a physics doctorate at Copenhagen University by tracing ancient climate changes in ice drilled from the Greenland ice sheet. He became Svensmark's chief collaborator and together they set about pinning down the effect of cosmic rays on clouds in more detail. They also found a friendlier place to work.

The surprising match to low clouds

On top of running a division of the Danish Meteorological Institute, Eigil Friis-Christensen was project scientist for Denmark's first satellite, Ørsted, designed to monitor the Earth's magnetic field. He was gathering a team of more than 60 research groups from sixteen countries. So he had little time to investigate the cosmic-ray connection much further in collaboration with Svensmark, although he continued to give lectures on the subject.

Towards the end of 1997, Friis-Christensen became director of the Danish Space Research Institute, which was later renamed the Danish National Space Center. The government wanted to widen the scope of the institute's activities, adding research on the Solar System to existing cosmic astronomy. New initiatives included studies of the Sun and its effect on the Earth's space environment, magnetic field and climate. In 1998 Friis-Christensen invited Svensmark and Nigel Marsh to join the space institute's staff.

The International Satellite Cloud Climatology Project issued a new series of data covering the period from July 1983 to September 1994. In their new lab, Marsh and Svensmark examined the data every which way, according to the heights of clouds and their positions around the globe. They compared the cloud variations month by month, in each area and at three different ranges of altitude, with the cosmic rays recorded at Climax, Colorado, which varied in response to changes in the Sun's behaviour. The work was time-consuming and often tedious, but by 2000 they could report a clear result: 'Surprisingly, the influence of solar variability is strongest in low clouds.'

In other words, it's the clouds that reach no more than about 3,000 metres above the ground, where the cosmic rays are always weakest, that react to the increases or decreases. Recall that NASA's Earth Radiation Budget Experiment had already identified these low clouds as being responsible for 60 per cent of the cooling of the Earth by clouds. Their identification as the chief players was therefore an important milestone in the quest for the link between cosmic rays and the climate. What matters is the intensity of the most energetic cosmic rays, because they are the only ones capable of reaching the lowest altitudes.

By a statistical test, the match between low clouds and cosmic rays averaged year-by-year scores of 92 out of a possible 100 per cent – a very good correlation by the norms of climate science. Against all expectation, the clouds at middle and high altitudes seem indifferent to the variations. The

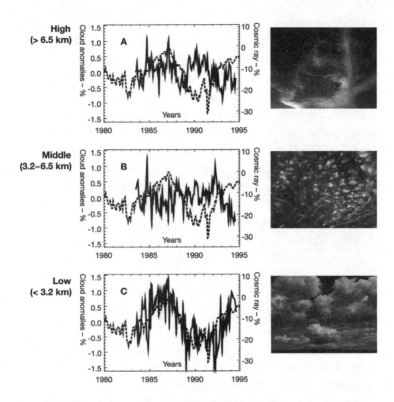

7. *Global variations of cloud cover at different levels in the atmosphere (solid line) are compared with the record of variations in cosmic-ray counts at the Climax station (broken line). While there is no match at the higher altitudes, there is a close correspondence between cosmic rays and clouds low in the atmosphere. (Graphs by N. Marsh and H. Svensmark)*

simplest explanation is that there are always plenty of cosmic rays at the higher altitudes, but variations are more noticeable at low altitudes where they are scarce – much as a shower of rain has a much more dramatic effect in a desert than in a rain forest. Moreover, the higher clouds consist of ice crystals, not liquid water, and may be formed by a different mechanism.

Large patches of the Pacific and Indian Oceans, and a region of the North Atlantic between Greenland and Scandinavia, show the strongest links between low cloud cover and cosmic rays. A more obvious geographical pattern emerged when Marsh and Svensmark's exhaustive analysis looked at the cloud-top temperatures. In this case, a belt encircles the globe, centred on the tropics, where the cloud behaviour follows the cosmic rays closely. The effect is emphatic over 30 per cent of the globe. When there are more cosmic rays, the cloud-tops of the low clouds become warmer and therefore radiate more heat into space, intensifying the cooling effect.

Why should the cloud-top temperatures respond to this influence from the stars? The most likely reason, Marsh and Svensmark suggested, is that there are more small specks in the air on which water droplets can condense. The clouds become foggier – with smaller droplets but more of them, and a smaller total of condensed water. As a result they are more transparent to heat from the surface. As now observed from satellites, at least two-thirds of the clouds over the oceans are in this strange regime.

Lines of clouds appearing in the wakes of ships at sea show just such an effect at work. It was verified by a research aircraft of the University of Washington that flew through clouds modified by two ship tracks in 1987. A satellite sees white streaks looking rather like the contrails seen behind aircraft. In reality they are much lower down, and they are just lines of brightening in pre-existing clouds, where the exhaust from a ship's funnel feeds many specks into the air.

Natural speck-making stimulated by cosmic rays could produce the world-wide warming of the low cloud tops. Colleagues sympathetic to the cosmic-ray scenario suggested that down-draughts of air carried specks formed at high altitudes to the low-cloud levels. But the Copenhagen pair disagreed. They suspected that the speck production was done in the lowest levels of the atmosphere, influenced by the relatively few cosmic rays that penetrate so far down. The next chapter tells how Svensmark gambled on that hypothesis, in planning a laboratory experiment.

When the Sun became busier

If cloudiness simply rose and fell every eleven years or so, in the rhythm of magnetic activity of the Sun that regulates the cosmic rays, any effect would even out and there would be no long-term influence on climate. But average cosmic-ray intensities declined markedly during the past 100 years, which implies a reduction in cloud cover and a warming of the Earth.

The temperature records show the Earth gradually warming by about 0.6 degrees Celsius overall during the 20th century. About half of the warming occurred before 1945, when the Sun was becoming more active and cosmic rays were diminishing, as indicated by rates of production of tell-tale atoms in the atmosphere. A period of pronounced cooling intervened in the 1960s and early 1970s, which accorded very well with a temporary weakening of the Sun's magnetic activity and a rise in the cosmic rays. After 1975, the upward trend in solar activity resumed for a while, the cosmic rays fell again, and the warming of the world resumed. That was the period when growing concern about carbon dioxide culminated in the creation of the Intergovernmental Panel on Climate Change.

Systematic measurements of the influx of cosmic rays went back only to 1937. But there were other ways to figure out what the cosmic rays were doing before then, so as to be able to gauge their effect during the 20th century as a whole. A remarkable revelation came in 1999 from Mike Lockwood and his group at the Rutherford Appleton Laboratory near Oxford. In interplanetary space, the Sun's magnetic field more than doubled in strength during the 20th century. This suggested an overall change very much in line with the temperature change.

Lockwood explained that his reckoning was made possible by a discovery by the European–US spacecraft Ulysses, that the solar field is equally strong in all directions. 'Nobody expected that, but it means we can use historical data from just one place, the Earth, to deduce a surprising change for the whole Sun.'

The historical data concerned magnetic storminess recorded at the Earth's surface (the *aa* index) which in turn is linked to the prevailing strength of the Sun's field. Spacecraft had directly measured an intensification of 40 per cent in the field strength since 1964, but Lockwood could infer a larger gain in the earlier part of the century, so that in total it amounted to 131 per cent, meaning that the field strength in 1995 was 2.3 times greater than in 1901.

During the more recent period of the field's strengthening, a detector at Huancayo in Peru showed a corresponding reduction in energetic cosmic rays, of the kind able to drive particles down to the low-cloud levels. From that change, Svensmark and Nigel Marsh were able reckon that the reduction of the relevant cosmic rays since the beginning of the century was 11 per cent. Translating that into the effect on clouds, they concluded that low-level cloudiness diminished by about 8.6 per cent as the Sun became busier. 'A crude estimate for the century trend in low cloud radiative forcing is a warming of 1.4 watts per square metre.'

That was a provocative figure to give, because the Intergovernmental Panel on Climate Change used the same 1.4 watts per square metre for the supposed global warming effect of all the carbon dioxide added to the air by human activity since the Industrial Revolution. Criticisms of the Copenhagen results therefore continued vigorously. One suggestion was that the variations in cloud cover studied by Svensmark had nothing to do with cosmic rays but were responses to volcanic eruptions or El Niño events. A simple lack of correspondence in time between volcanoes and cloud changes disposed of that possibility, but a match to the El Niños of 1987 and 1991 was quite good and was ruled out only by further analysis.

Other critics unwittingly used cloud data that the International Satellite Cloud Climatology Project had already discarded as unreliable. Many still imagined that cosmic-ray variations should have most effect on high clouds, because they are exposed to more intense cosmic rays. Ironically, when Jon Egill Kristjánsson and Jørn Kristiansen of Oslo University re-examined the proposed link between clouds and cosmic rays, they concluded that the only good connection concerned low clouds. So they dismissed it. With a different mind-set they might have been first to declare that it is precisely the low clouds that are affected by cosmic-ray variations.

Even when Nigel Marsh and Svensmark announced that conclusion in 2000, using a longer series of cloud data, some critics ignored it and continued to try to find fault with the original paper by Svensmark and Eigil Friis-Christensen. Although they could be rebutted one by one, the non-stop succession of antagonistic scientific papers served a purpose. Anyone who didn't want to take seriously the link between cosmic rays and climate could always say there were plenty of objections to the proposition. In 2001, the Intergovernmental Panel on Climate Change was still dismissive: 'The evidence for a cosmic ray impact on cloudiness remains unproven.'

Antarctica goes its own way

Experts were at that time waking up to the fact that temperature trends in Antarctica are repeatedly out of step with the changes occurring elsewhere. Details of this wayward behaviour give strong support to the proposition that clouds are the main drivers of the climate. Svensmark began searching for this confirmation back in 1996–97, when he was still at the Danish Meteorological Institute.

The satellite data from NASA's Earth Radiation Budget Experiment showed that clouds exert a warming effect in Antarctica, while their overall effect on most of the rest of the world is to cool it. Svensmark had already found the link between cosmic rays and clouds. If fewer clouds over the world as a whole could account for the general warming during the 20th century, then fewer clouds over Antarctica should have had a cooling effect. But reliable surface temperatures from the southern continent were hard to come by. And when he tried to calculate the effects of cloud cover, Svensmark did not fully appreciate how isolated Antarctica is, meteorologically speaking. Unable to predict a significant effect, he put the problem aside.

Windy carousels fence off Antarctica from the rest of the world's weather. That was what Svensmark had overlooked. Intense westerly winds in the Southern Ocean chasten sailors. They carry the wandering albatross on its regular journeys right around the continent and return it to its breeding ground. They also drive the great Circum-Antarctic Current, a rich feeding ground for whales. While connecting all the world's oceans at their southern extremities, the current separates Antarctica from tropical inflows like the Gulf Stream and the Kuroshio Current that warm the northern lands.

A similar vortex operates in the Antarctic stratosphere. Astronomers launched into it a balloon-borne telescope called Boomerang, in 1999, to find out how the Universe was built after the Big Bang. After travelling 8,000 kilometres in ten days, the telescope lived up to its name by landing just 50 kilometres from its starting point near Mount Erebus. The Antarctic polar vortex is more powerful and persistent than its counterpart around the North Pole.

While the Arctic climate tends to follow the rest of the world's, Antarctica might go its own way. After Svensmark's false start on the subject, evidence soon began to accumulate that it does so, in just the manner expected if the clouds are in charge. It came from teams drilling into the ice sheets at opposite ends of the Earth.

In 1999 Dorthe Dahl-Jensen of the Niels Bohr Institute in Copenhagen and her colleagues compared ice temperatures in the GRIP borehole in Greenland and the Law Dome borehole in Antarctica. Buried ice stores and insulates heat well enough to preserve its local temperature for several thousand years. A thermometer that samples the ice newly exposed during drilling operations, at different depths in the borehole, gives a direct record of the temperature at the time when each layer formed. When Dahl-Jensen compared north and south over the past six millennia, an alternation was plain. 'Antarctica has a tendency to warm up when Greenland is "cold" and to cool off when Greenland is "warm".'

These results were published in a glaciological journal. While Svensmark had heard about the results from Greenland, he missed the contrasting results from Antarctica until Dahl-Jensen's husband, Jørgen Peder Steffensen, mentioned them some years later. Svensmark commented that this was something he expected: 'Jørgen Peder did not seem receptive to my "Eureka". I kept thinking about the Antarctic issue, but other commitments got in the way.'

Dahl-Jensen's results showed that during the Little Ice Age of recent centuries, Greenland was distinctly cool, but Antarctica relatively warm. At another drill site in Antarctica, Siple Dome, Richard Alley and his colleagues from Penn State University found rare, distinctive layers where the ice had melted when it was on the surface during unusually warm summers. Changes in the frequency of melt events indicated variations in climate. In 2000, Alley's student Sarah Das announced a clear conclusion.

> Melting was most frequent between 300 and 450 years ago, with up to 8 per cent of years experiencing melt, most likely representing a period of elevated summer temperatures. This overlaps with the period of cool Northern Hemisphere temperatures often referred to as the Little Ice Age.

Das and Alley carried the ice-melt story back over ten millennia. They were struck by a period around 7,000 years ago, when no ice melting whatsoever occurred at Siple Dome for 2,000 years. While conditions were then particularly cold in Antarctica, they were unusually warm in Greenland. Ice of the same age from the GISP2 drilling site showed the greatest frequency of summer melting at any time during the past 10,000 years.

Other scientists found similar contrasts between Greenland and Antarctica, going even further back in time. Conventionally-minded climate scientists groped for explanations, typically involving changes in ocean currents. In

2001, Nicholas Shackleton of Cambridge expressed their bewilderment: 'Does a "polar see-saw" operate, with excess warmth flipping from one hemisphere to the other? What causes this enormous variability?'

For Svensmark there was no paradox. It was, after all, just what he hoped to see. And when, in 2005, he was at last able to give the matter fuller attention, he rejected the 'polar see-saw' tag as quite misleading. It suggested symmetry between the Northern and Southern Hemispheres, with a see-saw pivoted at the Equator.

In reality, the global climate is divided very unequally between isolated Antarctica and the rest of the world where winds and currents share the climate trends. Australasia, South Africa and South America and the oceans between them all have more in common with Eurasia and North America, in matters of climate change, than with nearby Antarctica. The pivot of any see-saw would be about 60 degrees south of the Equator.

A better name for the discrepancies is the Antarctic climate anomaly. And while others talked of time-lags between the contradictory climate changes, which you might expect if adjustments to the ocean currents were involved, Svensmark saw them as almost simultaneous. Whatever drives the different climatic responses in Antarctica and the rest of the world is still at work, on timescales of a few years at the most.

Temperature records since 1900 show an overall warming both globally and in Antarctica, but the steps along the way were mismatched. Big chills in the Antarctic region accompanied surges in global warmth in the 1920s and 1940s. Conversely, in the 1950s and 1960s, Antarctica warmed dramatically while the rest of the world experienced a temporary cooling. During the resumed warming of the world

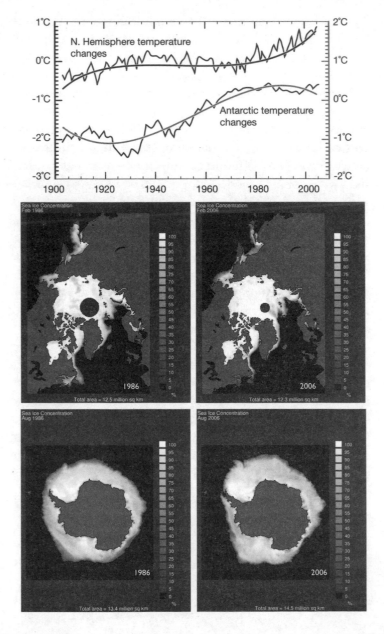

after 1970, temperatures in Antarctica levelled off. At one of the chief scientific stations, in Halley Bay, temperatures fell quite markedly.

Penguins know it's happening

How to account for the Antarctic climate anomaly? Which of the candidate drivers of climate change can explain it? Not carbon dioxide, because it spreads almost uniformly across the world, as far as the South Pole. Climate predictions based on increases in carbon dioxide suggest simultaneous and intense warming in the polar regions of both hemispheres, which has not occurred. A recent loss of high-altitude ozone over Antarctica, known as the ozone hole, could have helped to reduce surface temperatures there, because ozone acts as a greenhouse gas. But if, as suspected, the enlargement of the ozone hole has been due to recent releases of man-made fluorocarbons, that would have nothing to do with the Antarctic climate anomaly in historical and prehistoric times.

For astronomical reasons, the intensity of sunlight falling on Antarctica varies over thousands of years as the Earth's orbit around the Sun and its attitude in space gradually change. At present the Earth is closest to the Sun during the southern summer, but 10,000 years ago the northern summer had that advantage and Antarctica received weaker summer sunshine. That may help to explain why Dahl-Jensen's

8. [Left] *Climate-wise, Antarctica went its own way in the 20th century, as seen in the record of surface temperatures averaged over twelve years (lower graph). Whenever the Northern Hemisphere warmed (upper graph), Antarctica tended to cool, and vice versa. The charts for 1986 and 2006 show the concentration of sea-ice diminishing in the north and increasing in the south. This contrary behaviour is predicted if cosmic rays and clouds drive the changes of climate. (Surface temperature changes from NASA-GISS; ice charts from US National Snow and Ice Data Center)*

borehole temperatures for the Stone Age, 6,000 years ago, show Greenland relatively very warm, and Antarctica cold. But the astronomical changes (called the Milankovitch Effect) are far too slow to explain the rapid switches in northern and southern temperatures seen in the later part of the borehole record, and in the air temperatures measured in the past 100 years.

Cloudiness is the only forcing agent that directly predicts the Antarctic climate anomaly, without the need for any further process. When cloudiness decreases, the world will warm up and the southern continent will cool down. Increase the cloudiness and Antarctica will be warmer while the rest of the world is chillier. That is exactly the contrast seen. But why do the clouds affect the southern continent differently?

The snowfields of Antarctica create the whitest surface of the planet – brighter than the Arctic snow and whiter even than the cloud tops. As a result, the clouds absorb more energy from the Sun than the surface does in their absence, and they re-radiate heat towards the ground. Ground-based observations at the South Pole have confirmed the effects of Antarctic clouds seen by the satellites, as Michael Pavolonis of the University of Wisconsin-Madison and Jeffrey Key of the US National Environmental Satellite, Data, and Information Service reported in 2003: 'Clouds were found to have a warming effect on the surface of the Antarctic continent every month of the year.'

On the Greenland ice sheet too, surface warming by clouds has been known and measured for many years, and satellite measurements again show that a reduction of cloud cover has a chilling effect locally. As the Antarctic climate anomaly was discovered in the contrasting histories of the Greenland and Antarctic ice sheets, this northern effect might at first sight seem to rule out cloudiness as the driver. But the Greenland ice sheet is much smaller and its surface is not as shiny white as Antarctica's. Winds and ocean currents also couple Greenland's climate to that of the North Atlantic, and of the world as a whole, and they largely though not entirely override the local effect of warming by clouds.

Svensmark calculated the changes in surface temperature to be expected at different latitudes as a result of a small increase or decrease in the clouds, using the satellite data of the Earth Radiation Budget Experiment. When cloudiness increases by 4 per cent, temperatures should fall at the Equator by about 1 degree Celsius, and rise in Antarctica by about 0.5 degrees. For a 4 per cent reduction in cloudiness, the numerals remain the same, but minus becomes plus and plus, minus – with Antarctica cooling by about 0.5 degrees. Even with smaller changes in cloudiness, such figures are ample to explain the Antarctic climate anomaly seen in the 20th century.

A question remains. If the long-term warming trend in the world as a whole has been due to a reduction in cloudiness, why should Antarctica finish up warmer around 2000 than it was around 1900? Svensmark's answer is that, despite its isolation, the southern continent was able to share in the general warming because of a natural increase in water vapour in the atmosphere.

When the global atmosphere warms up, water evaporates more readily. As water vapour is the most important greenhouse gas, reflecting back to the surface some of the heat that would otherwise escape into space, it would have amplified the general warming provoked by a reduction in cloudiness during the 20th century. Extra water vapour also found its way into the air over Antarctica, Svensmark says, and its warming effect eventually overrode the cooling due to the loss of clouds. The Antarctic climate anomaly remains, in the alternating steps and setbacks along the graphs of rising temperatures.

In this analysis, completed in 2006, Svensmark's aim was to confirm the role of clouds as the primary drivers of the climate. In the computer models used by climate fore-casters, clouds play an essentially passive role, forming or dispersing in obedience to other forcing agents. But the clouds are really in charge, as Svensmark pointed out, because the contrary warmings and coolings in the southern continent say so: 'If changes in cloudiness drive the Earth's climate, the Antarctic climate anomaly is the exception that proves the rule.'

Despite problems in re-supplying the British Antarctic Survey's Halley station in 2002, when sea-ice barred the ship's way for the first time in 44 years, conventionally-minded climate scientists resisted any talk about an anom-alous Antarctic cooling. They pleaded a scarcity of decent temperature measurements. But migrating sea birds may rival human beings in assessing the trends, and they give a strong hint that the Antarctic climate anomaly still thrives.

The earlier appearance of migrating birds in spring is often cited as a demonstration of recent global warming in the northern lands. In 2006, after perusing records going back 55 years for Adélie penguins, Cape petrels, and other species breeding in East Antarctica, Christophe Barbraud and Henri Weimerskirch of the Centre d'Etudes Bio-logiques de Chizé in Villiers-en-Bois, France, noted that the sea-ice season had grown longer. The Antarctic birds were arriving at their spring-time nesting colonies nine days *later*, on average, than they did in the 1950s. But that was not a conclusion congenial for prevailing beliefs about the climate. Barbraud offered the suggestion that global warm-ing meant the birds were delayed by difficulty in finding food at sea.

Keep it simple, stupid!

A problem with climate science in general is that the system controlling events at the Earth's surface is elaborate enough for theorists to play endless games with it, moving ice, water, air and molecules around like chessmen, to explain any-thing they like. This was the case with the Antarctic climate anomaly, going further back in time. Although plain to see over the past 10,000 years, it's clearer still in the more drastic alternations in climate in the ice age, as it lurched between very cold Heinrich Events and much warmer Dansgaard-Oeschger episodes, as related in Chapter 1.

The temperature changes ascribed to those events relate to the Northern Hemisphere. In Antarctica they went the other way. Ice-drilling at the GISP2 site on the Greenland ice sheet and at Byrd in Antarctica supplied the best com-parisons. To make sure that they compared ice layers of the same age, investigators traced the ups and downs in the concentrations of methane gas from the air, trapped in bub-bles in the ice. Matching the methane wiggles at the two ends of the Earth made sure of the correspondences in age. The ice sheets also divulged their ancient temperatures by the counts of heavy oxygen atoms present in the ice itself. In 2001, Thomas Blunier at Princeton and Edward Brook at Washington State University were able to report on the major warming and cooling events recorded during the past 90,000 years.

Over this period, the onset of seven major millennial-scale warmings in Antarctica preceded the onset of

Greenland warmings by 1,500 to 3,000 years. In general, Antarctic temperatures increased gradually while Greenland temperatures were decreasing or constant, and the termination of Antarctic warming was apparently coincident with the onset of rapid warming in Greenland.

For an explanation, some experts imagined a reorganisation of ocean currents in the Atlantic, to carry heat southwards across the Equator instead of northwards as usual. But the climate anomaly operates nowadays on timescales of ten years or so, which is far too rapid for such a watery explanation. In any case, the ocean story is complicated and purely speculative.

When spinning hypotheses – so students are taught – it's fundamental in good science to use Occam's Razor, the principle of economy stated by the medieval sage William of Occam. One Latin rendering is *Entia non sunt multiplicanda praeter necessitatem*, which NASA's staff translated into American English as KISS, or 'Keep it simple, stupid!'

In other words, always prefer the simplest hypothesis or mechanism until it fails to do the job, and don't add any extra assumptions or embellishments unless they are unavoidable. In claiming the temperature swings in the ice age for his theory of cloud forcing in climate change, Svensmark uses Occam's Razor three times. He shaves away the complex mechanisms that have been invented just to explain the Antarctic climate anomaly, by pointing out that the unusual warming effect of clouds over the ice sheets of Antarctica accounts for it in the simplest possible way.

A second stroke of the razor leaves variations in cosmic rays, controlled by the manic-depressive Sun, as the simplest explanation for the variations in cloudiness. Thirdly, as the

Sun is clearly implicated in the contradictory warming and cooling events since the ice age, no other explanation need be invented for those that happened during the ice age. Occam's Razor will be worth remembering later in the book, when the very same mechanism of cosmic rays and clouds serves to explain climate change over millions and even billions of years. If that seems a greedy use of just one simple hypothesis, the answer is another Americanism: 'If it ain't broke don't fix it.'

Staying cool about carbon dioxide

The warming of the world during the 20th century is now explained twice over, by changes in the Sun's activity and by man-made greenhouse gases accumulating in the atmosphere, especially carbon dioxide. Either of these hypotheses, according to their fans, can on its own explain the increase in temperature of about six-tenths of a degree Celsius between 1900 and 2000. Both sides can't be right, of course, or the warming should have been twice as great.

For the sake of a quiet life, it might seem tempting to suggest that perhaps half the warming was due to the Sun and half to carbon dioxide. That may not do. Good science doesn't have to be fair or quiet, but it must aim to be correct and self-consistent. The greenhouse enthusiasts need to claim most of the 20th-century warming, to sustain their twin hypotheses that carbon dioxide is the main driver of climate change and that the world now faces a warming crisis. Half is not enough.

Nor would half of the 20th-century warming be enough for Svensmark's story. As later chapters will relate, the variable cosmic rays evidently caused climate changes in the past that were far more dramatic than anything seen since 1900. If a doubling of the Sun's magnetic field and the consequent reduction of cosmic rays did not take the lion's share of the 20th-century warming, it might be hard to account for bigger temperature swings at other times.

The longest of all systematic records of cosmic-ray intensities became available in 1998. Harjit Ahluwalia of the University of New Mexico retrieved old data from low-altitude stations set up by the cosmic-ray pioneer Scott Forbush in Cheltenham in Maryland and Fredericksburg in Virginia going back to 1937. Combining them with results from a similar instrument at Yakutsk in Siberia carried the series forward to 1994.

Using Ahluwalia's data, Svensmark compared the changes in cosmic rays with the changes of temperatures in the Northern Hemisphere. With fewer cosmic rays corresponding to fewer clouds and higher temperatures, the graphs danced through the decades, dipping just as expected between 1960 and 1975, and then climbing in company towards warmer times in the early 1990s.

Some scientists now say that the warming effect of carbon dioxide became clearly visible when the Sun's magnetic activity levelled off, along with cosmic-ray counts, from about 1985 onwards. It continued roughly flat overall until the time of writing (2006). Temperatures continued to rise, thereby ruling out (so it was said) any further contribution to climate change from any kind of solar effect, including variations in the cosmic rays.

The story about the Sun and recent temperatures is not as straightforward as that. Although the upward trend during the 20th century did indeed come to an end around 1985, solar activity did not fall away significantly in the following 25 years. The counts of cosmic rays continued to vary rhythmically as expected during each solar cycle, and the same rhythm can be detected in virtually every temperature record, superimposed as wiggles on any overall trend in temperature. This evidence of continued forcing of climate change by the Sun is particularly clear in the surface and sub-surface temperatures of the oceans, and in the air above the surface as measured by balloons and satellites.

The rise in temperature after 1985 was steepest in surface temperatures on land in the Northern Hemisphere. In other circumstances the trend was small or even non-existent, as if acknowledging that the Sun's contribution had levelled off. That was the case, for example, with the sub-surface water of the oceans down to 50 metres, which holds far more heat than the air does. The testimony of the water was that temperatures merely went up and down with the fall and rise in cosmic rays, as if global warming had stopped.

One puzzle for climate scientists is now to explain why the land surface north of the Equator seems to have warmed faster than the rest of the world by land, sea or air. If the meteorologists' records are reliable, there must be climate-changing processes at work in Northern Hemisphere land not present elsewhere. There are several candidates such as air pollution and changes in land use, in respect of which the oceans remain essentially unchanging.

Altogether more challenging is another conundrum. Why does the impact of increasing carbon dioxide and other man-made greenhouse gases appear to be much less than expected in most of the world? For example, they have not been able to override the cloud effect in Antarctica, where the area of sea-ice increased by 8 per cent between 1978 and 2005 – in a region long earmarked for rapid warming by the man-made greenhouse.

The extent of the variations in sub-surface temperatures in the ocean is entirely reasonable according to our story of cosmic rays and clouds. But the upward trend attributable to greenhouse gases over the past half-century is much smaller than you would expect if their effect has been correctly reckoned.

In this context it's worth recalling the opinion of the father of modern climate science, Hubert Lamb of the Climatic Research Unit in Norwich, writing in 1977.

> On balance, the effect of increased carbon dioxide on climate is almost certainly in the direction of warming but is probably much smaller than the estimates which have commonly been accepted.

Until the late 1980s, solar variability was widely recognised as the most likely candidate for driving climate change on timescales of centuries. Had the cosmic-ray story been known at that time, experts would have been much more confident about the Sun's effect. Supporters of the man-made greenhouse hypothesis would have had a hard time trying to get their show on the road, let alone to tell their tale of intolerable global warming to come. Now that the Sun reappears in its primacy, the burden of proof is back with the carbon-dioxide enthusiasts, to show what part of current climate change can be salvaged for their favoured mechanism.

In this spirit of déjà vu, the carbon-dioxide greenhouse goes back into a queue with other candidate contributors to climate change. These include fluctuations in the frequencies of major volcanic eruptions and the El Niño warming events, changes in the amount of dust and smoke in the air, variations in ozone, methane and other greenhouse gases, altered land use, and even, ironically, a general darkening of the land by the vegetation fertilised by all that extra carbon dioxide.

Satellite measurements of the effects of clouds on the

Earth's temperature allow Nigel Marsh and Svensmark to estimate a warming close to 0.6 degrees Celsius for the reduced cosmic rays and cloudiness between 1900 and 2000. The satellite data also help to confirm the role of clouds, in accounting for the Antarctic climate anomaly. In contrast, remarkable uncertainties hamper any definition of carbon-dioxide forcing by the greenhouse effect.

Different calculations disagree by a factor of ten, by giving anything from 0.5 to 5 degrees Celsius for the effect of a doubling of carbon dioxide in the air. And in the real world, the greenhouse scientists can lay claim only to an observed warming, with little hope of demonstrating that carbon dioxide, rather than any of the other agents mentioned, was really responsible for it. They have nothing to match the satellite and surface observations in Antarctica that support the cloud-forcing theory uniquely.

When asked to comment on the contribution from carbon dioxide to the recent warming, Svensmark has stayed cool. A sensible answer requires, in his opinion, not a quasi-political debate with each side trying to score points off the other, but a more exact scientific reckoning of the effect of extra carbon dioxide. Then it may be possible to match that effect to some part of the observed changes in climate. But the outcome, he thinks, may be good news for the planet.

When the greenhouse people save what they can from the warming record, after taking the Sun's contribution fully into account, the effect of carbon dioxide may well turn out to be quite small. If so, any global warming in the 21st century is likely to be much less than the typical predictions of 3 or 4 degrees Celsius.

For ten years after Svensmark and Eigil Friis-Christensen reasserted a strong role for the Sun by announcing the link between cosmic rays and cloud cover, in Birmingham in 1996, their opponents pooh-poohed the very idea that cosmic rays could play any part in cloud formation. They said that there was no physical mechanism to explain it. That line of defence is now taken from them, thanks to an experiment in a Copenhagen basement in 2005 that demonstrated exactly how exploded stars contribute to the aerial scenery of the clouds.

In a historical perspective, the experiment was just the latest step in a long struggle to understand what help water vapour needs if it is to make cloud droplets. That chapter of the story began in the 19th century.

4 Getting piggy over the stile

Clouds form when water vapour cools and condenses · The water vapour settles on specks floating in the air · The most important specks are sulphuric acid droplets · How these droplets appear was unexplained till now · An experiment shows how cosmic rays help them grow

In Victorian times, Britain led the world in industry and in the air pollution that went with it. London became notorious in the age of coal for the choking pea-soupers of November. Claude Monet painted the weird lights and shades produced by sunlight filtering through the fog at Westminster. In *Bleak House*, Charles Dickens saw a metaphor for litigation.

> Fog everywhere. Fog up the river, where it flows among green aits and meadows; fog down the river, where it rolls defiled among the tiers of shipping and the waterside pollutions of a great (and dirty) city. ... Chance people on the bridges peeping over the parapets into a

nether sky of fog, with fog all round them, as if they were up in a balloon and hanging in the misty clouds.

The smoke and sulphurous fumes pouring from industrial and domestic chimneys not only made the natural mists of autumn noxious and dirty. They intensified and prolonged the fog itself. In 1875 Paul-Jean Coulier, a French pharmacist, began experiments that John Aitken, a British engineer, unwittingly duplicated and then took much further. The starting point was succinctly described in a book, *The Wonderful Century*, written by the evolutionist and populariser of science, Alfred Russel Wallace.

> If a jet of steam is admitted into two large glass receivers – one filled with ordinary air, the other with air which has been filtered through cotton wool so as to keep back all particles of solid matter – the first will be instantly filled with condensed vapour in the usually cloudy form, while the other vessel will remain quite transparent.

Aitken wasn't surprised. He watched how materials switch to and fro between solid, liquid and gas. Very clean water is hard to freeze, even when cooled well below the freezing point. And when a solvent containing a salt or a molecular compound evaporates, and crystals should be forming, they are reluctant to do so without a small seed to grow on. In changes of state, as Aitken realised, hesitation and help-mates are the norm.

One use for his cloud-making in a bottle was to measure the dirtiness of urban air, by its efficiency in seeding the formation of water droplets. Here was ammunition for clean-air campaigners, although serious smoke-control did not come to London and other cities until the mid-20th century. Quite apart from that application of his work, Aitken made one of the top discoveries in the science of the weather.

Wanting to imitate natural cloud formation more closely, he let a jarful of cool air become saturated with water vapour. He then sucked some of the air out with a pump, so that what remained expanded and cooled. Then it was like moist air rising into cold layers of the atmosphere where the temperature drops below the dew-point and the air becomes supersaturated. With ordinary air, the jar promptly filled with a man-made cloud. If the air had been filtered it remained clear.

Aitken's conclusion was that the Earth could not produce clouds and rain if the water vapour were unable to condense into droplets on surfaces provided by specks drifting in the air. You need air saturated with water vapour twice over, at 100 per cent supersaturation or more, before water drops will form in clean air, with assistance from some *other source*. In the real world, a supersaturation of just 1 per cent is usually more than enough, thanks to an abundance of specks of a suitable size, called cloud condensation nuclei.

A by-product of this 19th-century line of research was a tool for atomic physicists: the cloud chamber. Charles Wilson, a Cambridge physicist usually known as C.T.R., wondered about the little water droplets that appeared even in clean air, given sufficient supersaturation produced by a sudden expansion of the air in a chamber. Suspecting that electric charges were the *other source* that promoted the condensation, he confirmed it with a beam of X-rays, which strip electrons from the air molecules to produce swarms of charges. When fired into his primitive cloud chamber, the X-rays filled it with a rain of droplets.

Later, Wilson found that individual sub-atomic particles

made tracks of droplets as they whizzed through the cloud chamber and left behind them a trail of charges. That caused a sensation. When he perfected a cloud chamber especially designed for particle-hunting, atomic physicists like Ernest Rutherford could scarcely contain their enthusiasm for the beautiful photographs that resulted. Cloud chambers figured in many investigations of cosmic rays in the early 20th century, including the first known piece of antimatter.

A lifelong fascination with clouds that inspired Wilson's experiments first began when he watched them come and go on a mountaintop in his native Scotland. Even while doing the work on sub-atomic particles that won him the Nobel Prize, meteorology remained his first love. And although he was never quite able to demonstrate it, Wilson was sure in his later years that cosmic rays must be involved in the weather. One of his ideas was that they affected lightning.

Svensmark was unaware of that long-forgotten aspect of Wilson's work. But when he first found that the Earth's cloudiness varies with the intensity of cosmic rays, memories sprang to mind of hands-on experience with a cloud chamber at his high school in Elsinore. He recalled, too, the photos he had seen as a student, of the droplet tracks left by cosmic rays. In some sense, he supposed, the Earth's atmosphere acts like a giant cloud chamber and responds to additional cosmic rays with increased condensation going into clouds.

That was an over-simplification, as Svensmark well knew. The thin and sketchy tracks left by all the cosmic rays, even if they could find highly supersaturated air, would be as nothing compared with the billion tons or so of water vapour that condense to make clouds every minute of the day. The cosmic rays had to amplify this natural action in some way. Perhaps they interfered with processes on a molecular and microscopic scale that create the cloud condensation nuclei, or made them more droplet-friendly. From the outset, Svensmark realised that this interplay, whatever it might be, would have to be tracked down in a careful laboratory experiment in the tradition of Aitken and Wilson.

The very idea of such an experiment met with the hostility that Svensmark had come to expect. When he was asked to talk at a meeting of the Royal Meteorological Society in London in 1999, the members lined up to criticise their guest – even in the tea break, when Svensmark found himself face-to-face with the most formidable person present, a cloud physicist who was a former president of the society. A film crew recorded the following dialogue.

EX-PRESIDENT: What is the point of doing that experiment?
SVENSMARK: Well there have been written several papers where they discuss, 'Where do the cloud condensation nuclei actually come from? How are they formed?'
EX-PRESIDENT: Oh, we know that!
SVENSMARK: No, it is not known.
EX-PRESIDENT: You should not argue with me on cloud physics!

A remark about words. Objects small enough to float in the air are technically *aerosols*. They are often called *particles*, but that can cause confusion when the sub-atomic particles of cosmic rays loom large in the story too. *Dust* is a reader-friendly word, but it suggests solid material, whereas most cloud condensation nuclei are minute drops of liquid. Hence our preference for *specks*.

The smell of a seabird's breakfast

Ever since Paul-Jean Coulier and John Aitken demonstrated the role of specks in cloud-making, a never-ending task for researchers has been to identify and count the various kinds of solid and liquid materials in the air. For medics and their gasping patients, toxic air pollution is still a big concern. Climatologists have more than one reason to be interested, because even specks that don't make clouds can steal heat from the Sun's rays. Laser beams, aircraft and satellites have all helped the speck-hunters to build up a rich picture.

Wind-blown dust from dry soil, deserts and beaches makes a big natural contribution. Farming in semi-arid and drought-prone areas increases the supply. That happened catastrophically in the Dust Bowl of the US Mid-West during a long drought in the 1930s. Similar events still occur routinely in Africa and Asia. When the world suffered a period of global cooling in the 1960s, some meteorologists who wanted to blame agricultural dust production called it the human volcano.

A similar story concerns soot from forest and grassland fires. Often these are natural ignitions by lightning or volcanoes. But deliberate burning of trees, grass and plant waste has been a common practice in land management since prehistoric times. Nowadays, in the dry season in South Asia, the burning of plant material and coal produces a brown haze that stretches from the Arabian Sea to the Bay of Bengal.

The cosmic dust of meteorites figures in the roll-call of tiny things floating in the air. So do the pollen grains that provoke hay fever, while bacteria and the spores of fungi also abound, up to remarkably high altitudes. And a never-ending series of chemical reactions going on in the air involves many different elements and compounds that finish up in specks. The haze that you often see over a forest on a sunny day is due to hydrocarbon vapours released by the trees. Sunlight converts them into smoggy materials like those that plague cities, produced from hydrocarbons in car exhausts.

Volcanoes fling out mineral ash, which falls out quite quickly, and also sulphurous gases that become converted into minute droplets of sulphuric acid and other chemical specks. Much of the sulphur from explosive volcanoes goes into the stratosphere, above the level of normal cloud formation, where it is slow to fall out and spreads around the world. The scary red sunset depicted in Edvard Munch's painting 'The Scream' was inspired by the explosion of Krakatau in Indonesia in 1883, which contaminated the atmosphere as far away as Norway.

A big eruption can cool the Earth's surface for a few years, by intercepting the Sun's rays and warming the stratosphere instead. After Mount Pinatubo in the Philippines exploded in 1991, laser beams showed the backscatter of light from the stratosphere increasing a hundredfold. It declined only gradually and was not back to normal until 1996. The amount of sulphur put into the stratosphere was said to be about 10 million tons.

The oceans act in this respect like a huge, non-stop, watery volcano. They release large amounts of sulphur into the lower air. It emerges as a vapour called dimethyl sulphide – a simple blend of two carbon atoms, six hydrogen, and one sulphur. The British chemist James Lovelock first discovered it emanating from the open sea, far from land, in the early 1970s. Microscopic plants drifting as plankton in the surface water, with names like dinoflagellates and prymesiophytes, are the source. When grazing creatures rupture their cells and microbes break down their contents, dimethyl sulphide comes off.

To our noses, the vapour smells of shoreline seaweed or cooked corn-on-the-cob. For many birds that live far from land – storm petrels for example – dimethyl sulphide means breakfast. When they scent it, they follow it to the most fertile patches of the ocean, where food is plentiful. The odour fades during the day as chemical action in the air, driven by solar rays, converts the dimethyl sulphide into minute droplets of sulphuric acid.

Similar chemistry makes nitric acid droplets, from nitrogen oxides produced by lightning strokes or released by microbes in the soil. Another form of nitrogen that comes from many living sources is ammonia. It likes to team up with sulphuric acid, forming specks of ammonium sulphate.

The need for replenishment

The inventory of airborne specks becomes much less confusing when you simply ask which are the most effective, weather-wise. Wind-blown dust affects the climate by blocking sunlight, but it's too coarse to provide cloud condensation nuclei. The same is true of pollen grains – even the finest of them all, which come from forget-me-nots.

On the other hand, ultra-fine specks made from vapours and gases abound in the air, often no bigger than a smallish protein molecule, just a few millionths of a millimetre wide. They're much too small to help make clouds. But if they manage to clump together into specks measuring around 100 millionths of a millimetre (or 100 nanometres) they become ideal cloud condensation nuclei.

Sulphuric acid droplets, which require a little water for their construction, are the most important worldwide. The chief source of sulphur over the continents is nowadays sulphur dioxide from human activity – especially the burning of fossil fuels. With rapid economic growth in developing countries, the outpourings may now approach 100 million tons of sulphur a year. But they are concentrated over industrialised regions, and even though the pollution can spread a few thousand kilometres downwind, that still leaves most of the world scarcely affected by man-made sulphur dioxide.

Over vast tracts of the open oceans, covering more than half the planet, cloud-making relies especially on sulphuric acid droplets made from dimethyl sulphide. Although the total tonnage of sulphur released may be less than half that of the man-made emissions on land, the oceanic sulphur participates in the weather over a much greater area. If you have to pick out the world's top natural source of cloud condensation nuclei, it's that smelly gas from the inconspicuous micro-plants of the loneliest seas.

Sea salt is sulphur's chief rival as a provider of cloud condensation nuclei over the oceans. Grains of sodium chloride of an appropriate size come from the fountains of fine spray thrown up by storm waves, especially in the Roaring Forties and Screaming Fifties in winter. They probably supply no more than about 10 per cent of the necessary specks, but they can compete for the available water vapour during the assembly of sulphuric acid droplets.

When updraughts in cumulus clouds carry a cloud's water droplets up into the cold regions of the air, they freeze into snowflakes and hail particles. Alternatively, at high altitudes, water vapour can skip the liquid stage and make ice crystals directly, as you see in the high, streaky cirrus clouds. In either case a different set of specks comes into play, to act as ice nuclei on which the water crystallises.

The ice nuclei have to fool the wandering water

molecules into treating them as existing ice grains looking for more recruits. Microscopic samples of the common mineral kaolinite, originating in clay, seem to be Nature's favourite ice nuclei. For human experiments in rain-making, the material of choice is silver iodide smoke. It encourages cold clouds to make ice grains, which then fall more readily than water droplets do. Whether natural or man-made, the snowflakes and hail particles usually melt on their way to the ground.

Sooner or later, the cloud-making specks of whatever description will be washed out of the air by rain, hail or snow, injected into the stratosphere by updraughts in the tallest thunderclouds, or dragged down slowly to the Earth's surface by gravity. They have to be continually replenished. The advent of better detectors in the 1990s, which could register ultra-fine specks just a few millionths of a millimetre wide, revealed the creation of swarms of new cloud condensation nuclei, in events called nucleation bursts.

At a forest laboratory at Hyytiälä near Helsinki, Markku Kulmala and his colleagues monitor the bursts routinely. At ten o'clock on a spring morning, for example, after the count of specks has fallen steadily during the night, it quite suddenly starts to climb. By midday the numbers can increase nearly tenfold. Then the count levels out, but the specks are still growing in size, in a process lasting several hours. At sunset the numbers begin to decline again.

These and other replenishments ensure that a litre of air in the cloud-forming regions of the atmosphere over land contains millions of cloud condensation nuclei. Even over the open oceans there are typically 100,000 per litre. That's why meteorologists were ready to imagine that there are always plenty of specks, and therefore no reason to think that cosmic rays could make any difference.

An ultra-fine harvest off Panama

As they went about their business in the late 1990s, weather experts exuded great confidence, whether they were forecasting rain and shine for tomorrow or the climate for AD 2100. Outsiders were not to know just how sketchy some of the most fundamental science of the atmosphere really was. Even among the meteorologists themselves, few were aware that a key part of the basic operations of clouds – the very engines of the weather – was beyond expert comprehension.

A conundrum arises about the chemistry of the atmosphere. When the cloud condensation nuclei that seed the formation of water droplets are themselves droplets of other vapours such as sulphuric acid, how do they form? Don't they need seeding too? Recall the old woman in the English folk tale trying to drive a pig back from the market.

> Fire, fire, burn stick! Stick won't beat dog! Dog won't bite pig! Piggy won't get over the stile and I shan't get home tonight.

To bring home the bacon in the form of sulphuric acid seeds for cloud-making, the standard account of the atmospheric chemists required brute force and plenty of time. The theory relied on high concentrations of sulphuric acid molecules in vapour form. These should recruit a few necessary water molecules and slowly club together in droplets, molecule by molecule, with no outside help.

The theory died a sudden death. Far too many specks turned up one day, in the wrong place. It happened over the Pacific Ocean, where scientists like to go to study the life cycle of the clouds, far from the confusions of man-made air pollution. An Orion naval patrol aircraft, adapted for research by the space agency NASA and fitted with the instruments needed to detect gases, vapours and small specks, made many flights among the clouds of the tropical Pacific.

Early one afternoon in 1996, the aircraft flew low over the Pacific waves south of Panama. Like a seabird, it was sniffing for dimethyl sulphide. A team of scientists led by Tony Clarke of the University of Hawaii chose this region because they hoped it would be particularly rich in marine life. They wanted to trace the chemical transformation of dimethyl sulphide in the air.

The pilot dropped to a height of 160 metres over the target area and, sure enough, the dimethyl sulphide showed up in the sensor in satisfying abundance. A clean wind blew in from the wide ocean to the west while the aircraft cruised at low altitude for an hour. Broken low clouds predominated in the weather, with occasional thundery showers.

The instruments showed the expected conversion of dimethyl sulphide in reactions involving water vapour and ultraviolet light from the Sun, first into sulphur dioxide gas and then into sulphuric acid vapour. The number of sulphuric acid molecules fluctuated quite a lot, but it remained far too low for them to club together, according to the prevailing theory.

A surprise came at two o'clock in the afternoon, when a detector on the aircraft encountered great numbers of ultra-fine specks. In two minutes the count shot up from nearly zero to more than 30 million per litre of air. The number of free sulphuric acid molecules measured at the same time remained low.

That burst of ultra-fine specks should simply not have been there, with the available concentrations of sulphuric acid. Unable to explain this precocious nucleation of the chief source of cloud condensation nuclei over half the globe, the weather scientists were left like a car mechanic who can't say what fires the spark plugs. Putting on a brave face, a preliminary report from NASA gave the unexpected findings a positive spin.

> What is quite clear is that this unique observation of a tropical nucleation event will provide a solid experimental foundation from which new theories can be tested.

When Clarke and his team cast around for an explanation, they wondered if ammonia coming from the ocean surface could have helped to speed up the formation of the ultrafine specks. A more far-out suggestion was that electric charges in the air, perhaps liberated by lightning flashes seen during the flight, might have encouraged the sulphuric acid and water molecules to come together.

That electrically charged molecules, atoms and electrons in the air, collectively called ions, might help to seed the formation of cloud condensation nuclei was not a new idea. It had been in circulation since the 1960s, with observations of ultra-fine specks that foreshadowed Clarke's, a few modest lab experiments and some preliminary hypotheses to back them up. An advocate in the 1980s was Frank Raes, a Belgian atmospheric chemist at the University of California in Los Angeles, who calculated that ion seeding of sulphuric acid micro-droplets was indeed feasible.

Open-minded theorists had ion seeding available on the bookshelf, so to speak, and they could not rule it out. It failed to fire their imaginations until they digested the implications of the discovery off Panama, made public in 1998. Then other atmospheric chemists began to realise that ion seeding might indeed spur piggy very quickly over the stile. And that would bring cosmic rays more persuasively into the story, because they, rather than strokes of lightning, are the main source of ions, both in the cloud-forming levels of the air and near the surface where the Orion flew.

Fangqun Yu and Richard Turco, also in Los Angeles, had studied the contrails that aircraft leave behind them across the sky. There too, in an aircraft's wake, cloud condensation nuclei form far more rapidly than expected by the traditional theory that wanted to piece them together molecule by molecule. Charged atoms and molecules produced by the burning fuel evidently help the specks to form and grow.

It was then a short step for Yu and Turco to acknowledge, in 2000, that the ions created by cosmic rays could assist in making cloud condensation nuclei – and therefore clouds. The presence of electric charges would encourage the molecules to come together at lower concentrations of sulphuric acid vapour than would be possible without them. The ions would then stabilise the resulting embryonic specks while they assembled into larger specks. Calculations accounted for the ultra-fine harvest off Panama.

Trying for an experiment

This timely intervention by the atmospheric chemists gratified no one more than the particle physicist Jasper Kirkby, at the CERN laboratory in Geneva. In December 1997, when Calder had given a lecture at CERN about Svensmark's detection of the link between cosmic rays and clouds, Kirkby was in the audience. His curiosity sparked, he took a collection of papers when setting off with his family to spend Christmas in Paris with his sister-in-law. During quiet times while the others were shopping, Kirkby studied the published science and became convinced that Svensmark's discovery was very interesting.

As the match between changes in cosmic rays and cloud

cover did not prove cause and effect, Kirkby wondered how the link might be tested by finding a mechanism that could initiate cloud formation. Although climate science is not the stuff of particle physics, cosmic rays certainly are – and particle physicists produce artificial cosmic rays in abundance from their accelerators. Amid the festivities in Paris, Kirkby found time to sketch an experiment. It would reproduce atmospheric and cloud conditions in a specially-designed chamber and measure the influence of a CERN particle beam.

This was a special opportunity for his esoteric branch of fundamental science to shine in environmental research, by addressing a possible cause of climate change. By feeding into the chamber controlled amounts of water vapour and traces of materials like sulphur dioxide, ammonia and nitric acid, the scientists could follow physical and chemical events in a series of carefully planned experimental runs, and see whether the incoming particle beam could affect the production of cloud condensation nuclei. From 'cosmics leaving outdoor droplets' Kirkby constructed a name for the experiment: CLOUD.

He set about assembling a team. Within two years he had recruited more than 50 atmospheric scientists, solar-terrestrial physicists and particle physicists from seventeen institutes in Europe and the USA. Svensmark was among them. He despaired of getting funds in Denmark for an experiment of his own and was glad to join Kirkby's troupe of experts.

So was Markku Kulmala of Helsinki, who had thrown Svensmark a lifeline in a sea of critics at Elsinore. Like some others in the team, he was not yet persuaded by Svensmark's results to suppose that cosmic rays participated directly in cloud formation. At that time he preferred to explain bursts of ultra-fine particles by reactions involving other molecules beside sulphuric acid and water vapour. But like everyone else he saw an opportunity to check the ion-seeding idea while carrying out unprecedented research on atmospheric chemistry.

Kirkby found a niche for CLOUD among the beam lines in an experimental hall of CERN's Proton Synchrotron. This machine would deliver a controllable number of high-speed particles to the cloud chamber half a metre wide, which was to be the centrepiece of the experiment. Team members from Helsinki, Missouri-Rolla and Vienna had favourable experience with cloud chambers, and CERN's engineers had built a big bubble chamber for particle-tracking using similar technology.

State-of-the-art instruments arrayed around the cloud chamber were to monitor the events provoked by the particle beam from the accelerator. Liquid droplets forming in the cloud chamber would announce their presence by scattering light. High-speed photography with a 3-D camera could exploit technology first developed for observing an eclipse of the Sun.

Atoms, molecules and ions of different kinds and masses present in the air would reveal themselves to several instruments. Three different kinds of mass spectrometers were to identify them by precise measurements of their molecular weights. Another instrument would gauge the mobility of the ions, which tells a lot about how they interact with the air and with other materials present in the experiment.

What the proposal lacked was adequate theoretical support from atmospheric chemists for a possible role for the cosmic rays, beyond the suggestions by Frank Raes dating from the 1980s. The more refined scenario from Fangqun Yu and Richard Turco in Los Angeles, accounting for those unexpected micro-specks off Panama, came at just the right time. By April 2000 the team had put together a detailed proposal. Its concluding words matched Svensmark's first thought.

> More than 100 years ago C.T.R. Wilson invented the cloud chamber to investigate weather phenomena. It evolved into a prime instrument for particle physics. Now the wheel of history turns and we go back to Wilson's concept to investigate the possibility that the Earth's atmosphere acts like one big cloud chamber that echoes the whims of the Sun.

When the proposal went out to two leading atmospheric scientists for review, the response was disappointing. A Nobel prize-winner scoffed at Svensmark's findings and felt obliged to draw CERN's attention to their use in an ongoing scientific-political debate about global warming. The team's public response was that this was not a scientific argument in favour of or against the proposal. A comment circulated privately questioned the logic of the objection.

> If the situation is as disagreeable as [the reviewer] describes it, then is it not all the more important to show that Svensmark's hypothesis is false?

The other reviewer took issue with technicalities in the proposal and questioned the ability of the experiment to simulate conditions in the real atmosphere. There, Kirkby's own cloud experts responded carefully to the issues point by point. They also gave assurances that the aim was not to prove that clouds *are* sensitive to cosmic rays, but only to see whether or not they *could* be.

The most telling technical objection was that an experimental run might be too limited in time. For embryonic specks to form and grow can take many hours. The loss of droplets sticking to the walls of the cloud chamber would end the run in about 24 hours. The team remedied that by combining two ancillary tanks in one large reaction chamber with more than 60 times the volume of the cloud chamber, and with Teflon walls. There, the chemical action could continue quietly for a few days or even a week.

CERN's own committee advising the director-general on the science programme demanded a clearer picture of how the experimental runs would proceed. The experiment was novel even to atmospheric physicists, let alone CERN's panel of particle physicists. Because he hoped that CLOUD would be approved before the end of 2000, the energetic Kirkby chivvied his experts into quickly producing a big addendum to the proposal. It described the new reaction chamber and detailed some initial experiments. A second addendum explained that, as so many years of experiments would be possible with CLOUD, it should be considered as the core of a semi-permanent atmospheric research facility in Geneva.

But shouldn't the cobbler stick to his last? Members of the CERN committee considering the proposal wondered whether a particle physics lab ought to get into atmospheric research at all. So they deferred making any decision. A long effort then followed, to build up a wave of support among atmospheric scientists.

The European Geophysical Society, the European Physical Society and the European Science Foundation co-sponsored a workshop in Geneva in 2001 to review 'ion–aerosol–cloud interactions' and discuss the experimental programme. It attracted 50 experts from around the world. On a show of hands they were divided equally between 'Yes' and 'Don't know' on the question 'Does cosmic-ray ionisation play a role in the climate?', but they were unanimous in supporting Kirkby's project.

Although this successful meeting raised spirits for a while, the bell tolled for CLOUD later that year. A vastly more expensive project was under way at CERN, to construct the world's most powerful accelerator, the Large Hadron Collider. It was stretching the multinational lab's budget to the limit and the directorate decided to bar any new experiments for the time being. That included even CLOUD, which was not expensive by the norms of high-energy physics.

Undaunted, Kirkby set out to persuade the Americans to provide the necessary accelerator for CLOUD. The best bet was the Stanford Linear Accelerator in California, where Kirkby had worked in the 1970s in the run-up to the discovery by Martin Perl of one of the Universe's most fundamental particles, the tau lepton. Perl himself was enthusiastic enough to join the CLOUD team. So did Fangqun Yu, who by then was at the State University of New York in Albany. Again the proposal had to deal with rather hostile reviewers, and the transatlantic ploy came to nothing in the end.

The project went on ice for three years, while the scientific case could develop and be more widely appreciated. Support for new experiments became possible again at CERN towards the end of 2004. Kirkby marshalled several of the big guns in his team for conversations with the most senior research management, before a meeting of the science programme committee in January 2005. There, Markku Kulmala from Helsinki made a persuasive presentation and the committee decided that CERN should provide the particle-beam facilities for CLOUD. In a message to Calder, Kirkby was jubilant.

The CERN side is now essentially taken care of. Provided our [national] funding is successful, we have a real CLOUD experiment and we can finally get to the physics. There are plenty more hurdles to cross, but the toughest one is behind us.

Seven years had passed since Kirkby first sketched the idea of the experiment. The formal proposal had been submitted to CERN nearly five years earlier. With good fortune, the team could expect to have their main experiment taking data by 2010.

The box in the basement

At the Danish National Space Center in the meantime, Svensmark and his colleagues had created and started running a more modest investigation of their own. Rather than waiting for an accelerator lab to condescend to provide a beam of particles that would liberate electric charges in an air sample, they let the natural cosmic rays raining on Copenhagen do the job for them. The experiment's name is SKY, which means *cloud* in Danish and is pretty apt in English too.

When the muons or heavy electrons, the most penetrating of charged cosmic-ray particles, hit the roof of the building on Juliana Maries Vej that houses the space centre, they take very little notice. They carry on down through floor after floor, via desks, computers, coffee cups and personnel. Before the muons disappear into the Earth's crust, some of them whiz through a big box of air standing in the basement, where they oblige Svensmark's team by knocking electrons out of the nitrogen and oxygen molecules and so creating ions.

SKY was conceived in 2000 when the news from CERN

was frustrating. It was to be a simpler way to start homing in on the processes in the air that make cloud condensation nuclei. The new calculations by Fangqun Yu and Richard Turco, explaining the surprising ultra-fine specks in the Pacific air, suggested to Svensmark that a relatively inexpensive set-up could look for such a process in the lab. It could be regarded as a pilot project that might be put together more quickly than CLOUD.

This was a new departure for Svensmark. Like Jasper Kirkby in Geneva, he was a physicist, not an atmospheric chemist. Moreover, as a theorist, he was not accustomed as Kirkby was to the patient pace of an experimentalist's life. Just to obtain a site for the box of air took time. A cleanroom in the basement, the one suitable place, was occupied by books that could be moved only by special permission of the university library. And as for funding, Svensmark could only hope he would get lucky.

Construction began in a tentative way with small grants from two private foundations. The prospects were at first so uncertain that SKY came last in the technicians' priorities. Work stopped many times. The Danish Natural Science Research Council, SNF, put SKY on a somewhat firmer footing with a grant of 600,000 kroner (roughly US$100,000) to be spread over three years, but it was still not nearly enough to finish setting up the experiment and gathering the necessary team as well. Even the funding from the Carlsberg Foundation that enabled Nigel Marsh to remain with Svensmark was coming to an end.

By 2002 the situation was grim. Some 50,000 kroner were needed urgently just to keep the project ticking over for the time being. Svensmark recalled that a leading industrialist had been very interested in his work, when heading a committee that awarded him the Energy-E2 Research Prize the year before. After many attempts, he at last got through by phone and started to explain his predicament. The industrialist at once sent a taxi to collect Svensmark, who found himself unshaven and in sandals in a roomful of people in suits. What they said stunned him. 'We were thinking of giving you 1,000,000 kroner in the first year, 500,000 the next, and 250,000 in the third year.'

That made a difference! Svensmark was able to keep Marsh in the team and to recruit an experimentalist from a physics lab in the Niels Bohr Institute. Jens Olaf Pepke Pedersen, an expert on collisions between fast particles and atoms, became a key colleague in getting the SKY experiment up and running. Full-scale operations would need still more funding, but the prospects became altogether rosier in 2003.

Danish parliamentarians can bypass the government's funding agencies, in order to diversify the support for special projects. By this route, Svensmark managed to get his own line in the national budget. Such support for supposedly wrong-headed research provoked outrage in the Danish media, from hard-line environmentalists and some scientists too. But it secured the project for the next four years with 12,000,000 kroner – twenty times the research council's grant for the SKY experiment.

Svensmark re-named his group the Center for Sun–Climate Research. Besides Marsh and Pepke Pedersen, the team grew to include Ulrik Uggerhøj, an atomic physicist at Aarhus University, and a PhD student, Martin Enghoff. With secure funding they could afford all the essential equipment, and get on with the experiment.

Historians of science looking back on this little saga may wonder why both Kirkby in Geneva and Svensmark in Copenhagen had such a white-knuckle ride to get approval and funding for separate projects costing just a few million US dollars. The world was spending billions of dollars a year on climate research at the time. Further food for retrospection will be the assertions by opponents, including some very eminent scientists, that they *knew* that the results of the experiments would be negative. Svensmark himself had no idea what a surprise was in store when, after a long period of testing, testing, and more testing, systematic experimental runs began shortly before Christmas 2004.

In the blink of an eye

Pipes, pumps, dials and electronic read-outs surrounding the 2-metre-high box of air gave the basement in Copenhagen the look of a ship's engine room. The impression was not altogether misleading because, judging by the quality of the air in the box, you could be in the middle of the Pacific Ocean and not a European city. Made of Mylar plastic lined with Teflon, and formally called the reaction chamber, the box held 7 cubic metres of ordinary air purified by passing through five different filters.

To check that nothing significant eluded the filters, the experimenters could fill the box with even purer air, made from bottled nitrogen and oxygen in the right proportion. In case nitrogen molecules played any chemical role in speck-making, argon sometimes replaced the nitrogen in the synthetic air. None of these variations in the air supply made any difference to the results. Ruling out nitrogen as a contributor eliminated a whole class of conceivable reactions involving positively charged ions. Instead, attention focused on the nimblest of negative ions, the electrons.

The temperature and humidity of the air were under control, and measured traces of sulphur dioxide and ozone went into the reaction chamber. Seven ultraviolet lamps, running either continuously or in a burst lasting ten minutes, played the Sun's part in powering chemical reactions. A detector of ultra-fine specks would show the products of the chemistry.

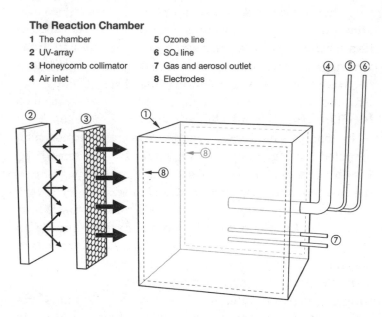

The Reaction Chamber

1 The chamber
2 UV-array
3 Honeycomb collimator
4 Air inlet

5 Ozone line
6 SO₂ line
7 Gas and aerosol outlet
8 Electrodes

9. *In the* SKY *experiment at the Danish National Space Center, cosmic rays coming through the roof entered a plastic box containing 7 cubic metres of purified air together with traces of sulphur dioxide (SO_2) and ozone – as found in unpolluted air in the natural environment. The amount of water vapour was also controlled. Ultraviolet (UV) lamps created sulphuric acid, which then joined with water molecules to make vast numbers of clusters. The production was less when a high voltage between electrodes swept away electrons set free by the cosmic rays, and greater when gamma rays increased the supply of electrons.*

Experiments starting with bursts of ultraviolet reproduced the creation of ultra-fine specks, as discovered occurring naturally over the Pacific. The ultraviolet rays provoked the rapid manufacture of sulphuric acid. Although there were far fewer molecules than required in the old brute-force theory of droplet formation, the sulphuric acid rapidly gathered into clumps.

Newly formed specks began to appear in SKY's detector after a delay of only ten minutes or less. In a typical case, they reached a peak count of about 2,000 per litre within the next quarter of an hour, even though the walls of the chamber were continually mopping up the specks. Taking that loss into account, the cumulative production ran into tens of millions per litre, matching the Pacific observations.

In that general sense, everything worked better than expected. But the course of events, from one experimental run to the next, made it a cliff-hanger. The starring role in the chemical drama unfolding in the box of air was supposedly taken by the cosmic rays coming in through the ceiling, exiting through the floor, and leaving charged particles in their wake. The experimenters were taken aback when they began to verify that that was the case.

In the very conception of SKY, Svensmark wanted a simple Yes or No to the question of whether the ions produced by the cosmic rays really were seeding the specks. To get the answer, he arranged that a powerful electric field could be switched on to sweep all of the charged particles from the box of air. It would remove them in a second. According to the available theories, the charged particles needed about 80 seconds to have a significant effect. So if they were really doing the work, the specks should not form. Svensmark later recalled what happened.

So it was an evening in the lab and everybody related to the project was gathered. The experiment was done with the electric field on, and the ultimate test of ion-induced nucleation was on its way – at least so we thought. But after 10 minutes the whole chamber was filled with ultra-fine particles, just as before. It was a very strange moment. Was it the end of the whole idea?

The first reaction was to check everything. Were the sulphuric acid concentrations properly measured? Was the ultraviolet system well calibrated? Would a honeycomb mask, obtained from the aeronautical industry and carefully painted, make the ultraviolet rays more homogenous? Everyone was on edge and tempers flared if something was not exactly to specification. The weeks passed in technical work before the team tried again.

Alas, the electric field still made no difference. Success or failure then depended on finding an explanation quickly. Svensmark wondered if the electrons – the lightweight negatively charged particles knocked out of the ordinary air molecules by the cosmic rays – might do their seeding work much faster than he or anyone else had ever imagined possible. That could be the case if the electrons jumped, leaving one embryonic sulphuric acid droplet to start another, like a teacher organising a crowd of children into teams.

If so, one electron could have a large effect in less than a second, before the electric field swept it away. Instead of trying to remove the ions, perhaps the team should create more, and see whether that increased the number of specks. Gamma rays could do the job, but the only radioactive sources that the team had ready to hand made little difference when put on the outside of the chamber walls. When inserted right into the box, in a tube, the gamma-ray sources provoked a strong nucleation of specks, which was encouraging.

A surprising effect showed up by chance a few days later.

Martin Enghoff, the PhD student, and Joseph Polny, one of the engineers, noticed that soon after they put the radioactive sources into the box, the ultra-fine speck detector started showing large counts. That was before the ultraviolet lamps were switched on, which were supposed to convert the sulphur dioxide into sulphuric acid vapour. Evidently the process could accomplish that initial chemistry very nicely, thank you, without the aid of ultraviolet light.

Although the first results with increased ionisation dispelled the earlier gloom, they were too improvised for formal tests. Another five weeks passed in other kinds of experimental runs, while waiting for better gamma-ray sources to come from Belgium, which would work more evenly throughout the box. Then detailed experiments with increased ionisation could begin.

They showed very clearly that the greater the number of charged particles set free in the air, the higher was the production of ultra-fine specks. To double the count of specks needed a fourfold increase in the number of ions. (In other words the productivity goes with the square root of the density of ions.) That means any variations in the cosmic rays would have most effect on speck production when the overall intensity was relatively weak.

So the ion seeding was real after all, and the implication was that the electrons truly did work in the blink of an eye, before the electric field swept them away. The team amassed a coherent set of results during six months of experimental runs of very different kinds, using the ultraviolet lamps in bursts or continuously. When confident also that he could account for the results theoretically, Svensmark returned to his original idea of suppressing the activity with the electric field.

He estimated that the jumping electrons would need about a fifth of a second to seed the sulphuric acid clusters. A more powerful electric field could test that by clearing the air more rapidly. Until then the limit had been 10,000 volts across the chamber. Near the end of June 2005, the team tried 20,000 volts and, encouragingly, the peak count of ultra-fine specks was down by half.

Next day they connected up a 50,000-volt generator that Ulrik Uggerhøj had found in Aarhus. When the voltage had risen past the 40,000 mark, a spark went though the box with a thunderclap. The electromagnetic pulse knocked out the electronics and one of the flowmeters. As the team hurried to put the system back in order, Svensmark expressed some amusement: 'With sparks and explosions it feels like real science.'

When they tried again on day three, limiting themselves to 40,000 volts, there was a pause – just a longer fuse, as Svensmark put it – and then the same thing happened again. Sadly the damage to the system and instruments was more severe, and would take three months to make good. Without the opportunity for any more experimental runs for a while, it was time to settle down and describe the results in a report for publication in a scientific journal.

Electrons seed the seeds

Fortunately, the data already amassed were singing to them, and Svensmark and his team could listen to the tune and try to understand it. The production of ultra-fine specks was far too fast according to previous ideas – even by the latest theory of Fangqun Yu and Richard Turco. A completely new mechanism for their nucleation was needed.

While the experiments were still going on, Svensmark developed a mathematical description of all the events *after* the first specks appeared in the instruments. Run in a computer, it simulated the results very well. The same maths run backwards, as it were, also offered a persuasive picture of what should be happening at sizes smaller than 3 millionths of a millimetre, *before* the instruments could pick them up.

From the sequence and pace of events, Svensmark saw that the action starts prematurely. Injection of the sulphur dioxide and ozone into the chamber occurs about an hour before the Sun is switched on, as simulated by the ultra-violet lamps. During that hour, clusters of molecules must form, much smaller even than the ultra-fine specks and therefore undetectable by the available instruments. Production of sulphuric acid proceeds without the aid of ultraviolet rays – as Martin Enghoff and Joseph Polny had confirmed by chance when handling the gamma-ray sources.

Electrons are the key players. Attached to an oxygen molecule, a single electron is enough to make it attractive to water molecules. Several gather around, making a water cluster. Activated by ozone and supplied with sulphur dioxide, the water cluster becomes a production centre where sulphuric acid can be manufactured – and where it can accumulate. Out goes the old picture of sulphuric acid molecules first being made with the aid of ultraviolet light, and then slowly clubbing together as an afterthought. Instead they are born into molecular clusters – at least in this very first stage of speck-making.

To begin with, the electron is the glue still holding everything together. But when the cluster has stockpiled a few sulphuric acid molecules, and is still very small, it becomes stable on its own account. Then the electron can move on, find another oxygen molecule, and start marshalling another cluster. So it acts as a catalyst, promoting the chemistry without expending itself.

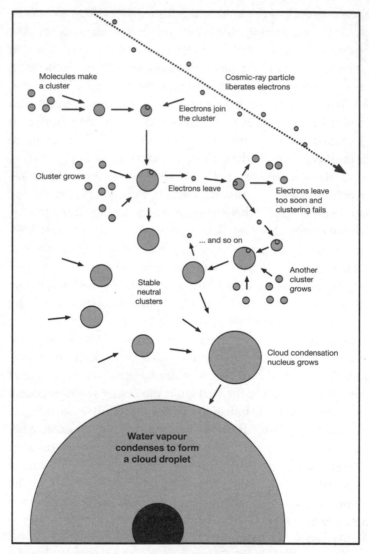

10. *The very rapid action of cosmic rays in creating the building blocks of cloud condensation nuclei, on which water droplets form, depends on busy electrons marshalling molecules together.*

The process is very fast, and with many electrons at work in the SKY box the number of molecular clusters reaches millions per litre before the ultraviolet lamps come on. When that happens, and sulphuric acid becomes much more abundant, the pre-existing clusters are ready to grab it. By the time they have amassed about 70 sulphuric acid molecules each, they have increased in diameter from 1 to 3 millionths of a millimetre (1–3 nanometres) and become identifiable as the ultra-fine specks.

If the new theory correctly explains the events in the reaction chamber, and if SKY is a realistic model of the atmosphere, then the same process must occur in the air above our heads. The ultra-fine specks will grow into full-sized cloud condensation nuclei, and seed the formation of our everyday clouds. And the answer to the conundrum about what seeds the seeds, what nucleates the nuclei, or what gets piggy over the stile, is electrons – liberated in the air by cosmic rays.

The experiment was finished by the summer of 2005 and written up in a scientific paper. The team then ran into long delays as one leading scientific journal after another rejected the report, for various reasons that did not reflect adversely on its technical merits. Especially frustrating was the rule that journals often enforce against premature public disclosure, which meant that nothing could be said openly about the experimental results until publication day. More than a year passed by without the news of the experimental results being known to any but a small circle of colleagues.

At last a prestigious London journal, *Proceedings of the Royal Society*, accepted the paper, entitled 'Experimental evidence for the role of ions in particle nucleation under atmospheric conditions'. Although publication of the printed issue of the journal was not due until 2007, the journal released it online in October 2006. Notices to the press went out from the Royal Society and from the Danish National Space Center, where Eigil Friis-Christensen, the director, added a comment.

> Many climate scientists have considered the linkages from cosmic rays to clouds to climate as unproven. Some said there was no conceivable way in which cosmic rays could influence cloud cover. The SKY experiment now shows how they do so, and should help to put the cosmic-ray connection firmly onto the agenda of international climate research.

Leaving aside all the personal and technical hassle of 1996–2006, a précis of the story till now goes like this. Weather satellites showed the Earth's cloud cover varying rhythmically over the years in accordance with the changing spottiness of the Sun – more precisely with the varying effect of the solar wind, which regulates the number of cosmic rays from the stars that reach the Earth. That prompted experiments in atmospheric chemistry. They revealed how electrons set free by the cosmic rays might catalyse the clubbing together of sulphuric acid molecules, the most important source of cloud condensation nuclei.

Although still open to more elaborate lab experiments like CLOUD at CERN, and to probings of the real atmosphere by aircraft, the chain of explanation from the stars to the clouds to the climate is now essentially complete. The SKY experiment successfully reproduced conditions low in the atmosphere where alterations in cosmic-ray intensities produce clear changes in cloudiness. The results give greater confidence to anyone wanting to examine what role the ever-changing cosmic rays have played in the ever-changing

climate of our planet. That is the theme of the chapters that
follow. While you can thank your lucky stars for the clouds
that water the world, be aware too of their power to chill.

5 The dinosaurs' guide to the Galaxy

The climate changes rhythmically over millions of years · Icy times occur during visits to the Milky Way's bright arms · Climate influenced evolution – e.g. in the origin of birds · Warming by carbon dioxide may be less than advertised · Now climate data give facts and figures about the Galaxy

Within living memory the farmers on the island of Møns, which lies in the Baltic Sea 80 kilometres south of Copenhagen, left out offerings of hay to protect their crops from the horse belonging to the King of the Cliffs. Supposedly a successor of the Norse god Odin, creator of the world, Klintekongen was reputed to be seen in the form of a bird, protecting the island from invaders. His home was a cave in Møns Klint, the most impressive sea cliffs of Denmark.

The soaring white chalk of Møns Klint tells a tale to geologists which rivals any folklore, about changes of climate that fashioned the modern world. The chalk itself was made around 70 million years ago, when tyrannosaurs and other giant reptiles still dominated the world. Conditions were so

i, ii. When ice melted in the pass of Schnidejoch in the Swiss Alps in the hot summer of 2003, it revealed a long-forgotten route for travellers. Among hundreds of objects found by archaeologists was this shoe from the late Stone Age. During the past 5,000 years the pass was open at four intervals as warm as today, telling of an ever-changing climate that followed the whims of the Sun. (Archäologischer Dienst des Kantons Bern)

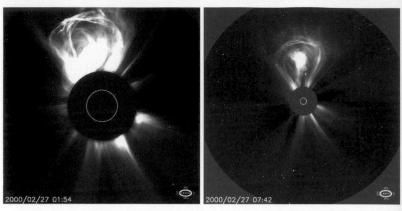

iii, iv. Resembling a gigantic light-bulb, a vast mass of gas is blasted from the Sun by an explosion in its atmosphere. The size of the visible Sun is shown by the white rings on the masking discs. Coronal mass ejections like this help to repel cosmic rays coming from the Galaxy. (NASA and European Space Agency, SOHO spacecraft)

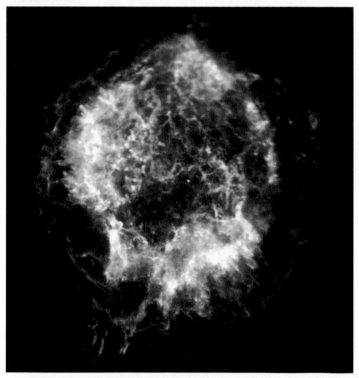

v. Wispy blue lines of energetic X-ray emissions show cosmic rays beginning to stir in the remnant of an exploded star. Cassiopaeia A shown here is the youngest known supernova remnant in our Galaxy, created when a giant star blew up in the late 17th century. (NASA/CXC/UMass Amherst/M.D. Stage et al., Chandra spacecraft)

vi. Spectacular displays of auroras in the upper air, like this one seen over Finland in late October 2003, are frequent evidence that the Sun toys with the Earth's surroundings like a playful tiger. For anyone who doubts the Sun's ability to influence the Earth's climate, an aurora should be a reminder of its awesome powers. (© Pekka Parviainen, www. polarimage.fi)

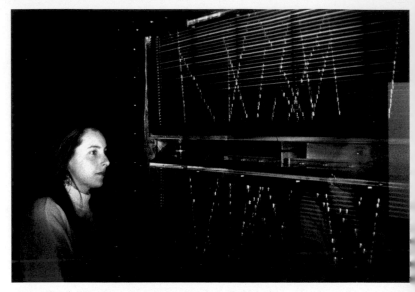

vii. Tracks of natural cosmic rays coming through the ceiling are made visible by sparks in a detector at a public exhibition at the CERN laboratory in Geneva. No one notices this ceaseless rain of high-energy particles riddling our environment and our bodies, but the clouds do. (© CERN)

viii. Continuous observation of the world's clouds by satellites made possible the discovery of their links to cosmic rays. This is a composite for one day – 18 January 2005 – using infrared images from Europe's Meteosat and US GOES satellites, stationed over the Equator. Between them they supervise the cloudiness of the whole globe except for the polar regions. (© 2007 EUMETSAT)

ix. At the summit of the ice sheet in the 1990s, the Greenland Icecore Project (GRIP) drilled deep into the underlying layers of ice to recover many kinds of information about past changes of climate and their causes. It was one of several such projects on the world's ice sheets. This photo shows the GRIP main dome in 1990, a heated facility containing kitchen, dining room, bathrooms, communications and sleeping quarters for ten to twenty people. (Ice and Climate Research Group, Niels Bohr Institute, University of Copenhagen)

x. By lowering a thermometer deep into a borehole in the ice, scientists could take the temperature of successive layers of ice that formed thousands of years ago. Similar measurements in Greenland and Antarctica revealed contradictory climatic trends at opposite ends of the Earth. The photo shows temperature logging of a deep borehole in 2004. Each measurement for every ten metres had an accuracy of twelve-thousandths of a degree Celsius. (Ice and Climate Research Group, Niels Bohr Institute, University of Copenhagen)

xi. Henrik Svensmark and Jens Olaf Pepke Pedersen are here seen at work on the SKY experiment in the basement of the Danish National Space Center in Copenhagen in 2005. It detected the effect of cosmic rays coming through the ceiling on trace gases in otherwise very pure air, and revealed the chemical link between cosmic rays and cloud formation. (Lars Oxfeldt Mortensen, Mortensen Film, Copenhagen)

xii. Low clouds that cover huge areas of the world's oceans are the chief regulators of the climate. Their extent varies with the intensity of cosmic rays penetrating to low altitudes, which originate mainly from the most energetic particles thrown out by exploded stars. (© Margaret Worrall) xiii. This artist's impression of the Milky Way Galaxy, based on the latest astronomy, shows the suburban location of the Sun and its planets amid the outlying spiral arms where bright, short-lived stars are concentrated. While the Sun orbits around the centre of the Galaxy, the spiral pattern rotates like a catherine wheel, but at a different rate. As a result, the Earth experiences high cosmic rays and cold climates during the Sun's passages through the spiral arms. (NASA/JPL-Caltech/R. Hurt (SSC/Caltech))

xiv. Enormous jets of hot gas pour from the starburst galaxy M82, produced by the explosions of many massive stars during a boom in star formation. A close encounter with another galaxy, M81, provoked the starburst. When similar though less spectacular events occurred in our own Milky Way Galaxy, they drenched the Earth with enough cosmic rays to make it freeze over. (Mark Westmoquette (University College London), Jay Gallagher (University of Wisconsin-Madison), Linda Smith (University College London), WIYN/ NSF, NASA/ESA)

xv. Tropical shores must have looked like this during Snowball Earth episodes around 2,300 million and 700 million years ago, by the evidence of stones dropped from glaciers and icebergs near the Equator. The episodes coincided with starburst events in the Milky Way. Here the Moon is rising over Graham Land, Antarctica. (British Antarctic Survey/Andy M. Smith)

xvi. The Pleiades or Seven Sisters is one of the nearby clusters of hot, short-lived stars that may have subjected the Earth to climate-chilling bursts of cosmic rays during the past few million years, when now-missing members blew up. The brightest survivors are visible to the naked eye and there are hundreds of other stars in the cluster. The nebulous glow is real and due to dust from now-vanished stars, but the crosses and rings are produced by the telescope – the Schmidt survey instrument at Palomar, California. (NASA, ESA and AURA/Caltech) warm that even the polar regions were free of ice and Antarctica was an abode for dinosaurs. The sea-level was very high. Calcium carbonate plates that provided coats for billions of microscopic algae accumulated on the sea floor when their owners perished. In the Baltic region they consolidated into chalk 100 metres thick.

The same thing happened on an even vaster scale in other parts of the world and gave the geological period its name, the Cretaceous, which means chalky. Not far away, in southern England, the thick chalk was later uplifted and eroded where it stood, to make the White Cliffs of Dover. There, the layers of chalk remain to this day stacked as neatly as when they first accumulated over many millions of years on the Cretaceous sea-bed.

The scene at Møns Klint is so different from Dover that it provoked intense debate in the 19th century. The Danish geologist Christopher Puggaard expressed his consternation at what he found there, in a report published in 1851.

> The strata of chalk are twisted, curved, and bent in all ways, in the form of an S or a Z, in a semi-circle or stirrup shape, or, again, cut by chasms, forming enormous faults, and interlaced in the most extraordinary fashion. About the middle of the scarp, at a place called Dronningestol, the confusion attains its maximum, and there the cliff rises to its highest point. ... The dip of the strata also varies greatly and changes continually, in some places passing abruptly from a horizontal position to a vertical one.

There followed half a century of wrangling between geologists who favoured different explanations for Møns Klint – and for many other less spectacular examples of chalk displacement found in north-west Europe. Some, including Puggaard himself, spoke of subsidence or erosion of underlying rocks, so that the chalk and a coating of younger deposits collapsed, like a ceiling falling down. Others suspected that ice movements were responsible for the scrambled strata.

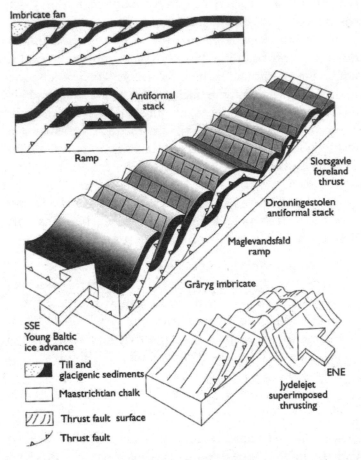

11. *From hothouse to icehouse: chalk laid down when the world was very warm was later bulldozed by a glacier to make the weird stacks of chalk mixed with glacier debris seen in the cliffs of Møns Klint in Denmark. (S.A.S. Pedersen, Geological Survey of Denmark and Greenland)*

Nowadays, the picture is much clearer. During the most recent ice age, beginning about 70,000 years ago, a big glacier advanced westwards across the region that is now the Baltic Sea. The snout of the glacier scraped up the chalk like a bulldozer, in two dozen flakes of chalk, each about 100 metres thick. It pushed them higgledy-piggledy ahead of it, until the ice ceased its advance. When the big thaw came, the island of Møns remained as a 'terminal moraine' where the glacier had dumped its load.

So Møns Klint is a product of two contrasting states of the world, nicknamed the hothouse and the icehouse. In the first, the chalk-making organisms flourished in balmy waters. In the second, their burial ground was torn apart by a moving mountain of ice. The transition was emphatic, but not sudden.

Temperatures were falling markedly around 50 million years ago. By 30 million years long-lasting ice sheets were in place on Antarctica. Really cold conditions in the North Atlantic region set in 2.75 million years ago. Since then the world has been in the icehouse mode, with glaciers and ice sheets providing an ever-present part of the scenery.

Experts try to account for that big change in climate in various ways. The continents were wandering about as usual, and changes in geography included the departure of Australia from Antarctica, leaving it isolated at the South Pole. Made lonelier still by the Circum-Antarctic Current that cuts it off from any warm influx of ocean water, Antarctica became a perfect platform for accumulating ice sheets. The collision of India with Asia pushed the Himalayas and Tibet high into the air and created a pool of coldness in the tropics. Another suggestion is that a fall in the amount of carbon dioxide gas in the atmosphere might have been responsible for the cooling.

An astrophysicist at the Racah Institute of Physics in Jerusalem, Nir Shaviv, has offered a quite different explanation for the switch, from the hothouse that bred the Møns chalk to the icehouse that trashed it. In his view, the answer to the puzzle lies in the Milky Way. Especially, he would say, in a very bright region called the Sagittarius-Carina Arm, which is best seen in the Southern Hemisphere on a winter's evening.

About 60 million years ago, the Sun with the Earth in company encountered that region, which was populated then as it is now by bright, short-lived stars. The Solar System came from the far side of the bright arm, as we see the Milky Way now. It emerged on the near side about 30 million years ago. There, the number of exploding stars was at a peak, and so was the intensity of cosmic rays generated by them.

Shaviv adopted the Danish findings about the climatic effect of cosmic rays, and their capacity to chill the world by increasing the low cloud cover. In this interpretation, global temperatures fell between 60 and 30 million years ago, and Antarctica acquired its ice sheet. As the Sagittarius-Carina Arm receded, the cooling hesitated, and it would have reversed if the wanderers through the Galaxy had not run into an extra fragment of concentrated bright stars, called the Orion Arm. That is where we are now, still deep in the icehouse, with the present relatively warm interlude between ice ages merely a respite from glaciations of the kind that churned up the Møns chalk.

Published in 2002, Shaviv's analysis accounted not only for the most recent hothouse-to-icehouse transition but altogether for four major chilling events since animals first became conspicuous on the planet, a little over 500 million years ago.

Everything said until now in this book, about cosmic rays and climate, has concerned recent changes, on geological and astronomical timescales so short that the influx of cosmic rays from the Galaxy towards the Solar System has scarcely changed. Variations in the Sun's behaviour have been the primary reason for changes in the intensity of cosmic rays reaching the lowest levels of the Earth's atmosphere over the last 100,000 years. As we move with the Sun and the Earth into wider realms of time and space – across millions of years and thousands of light-years – larger and longer-lasting changes occur in the influx of cosmic rays.

A message in the meteorites

Although you need a telescope to see any of them properly, spiral galaxies are among the most beautiful objects in the sky. These swarms of many billions of stars are so organised that the brightest and bluest stars are mainly dotted along gracefully curved arms that radiate out from a central ball or bar of generally older and redder stars. Gravity flattens a spiral galaxy, and seen edge-on it bulges in the middle like a fried egg.

Because we're right inside it, the Galaxy where we live appears simply as a band of light around the sky – named the Milky Way long before it was recognised as an 'island universe' similar to many distant objects scattered across the night sky. Not until the 1950s, when a Dutch radio telescope charted the distribution of hydrogen gas, were astronomers able to say with confidence that the Milky Way Galaxy is a spiral, like the Andromeda Galaxy, the Whirlpool and many others.

The force of gravity acting between the stars generates waves of denser and skimpier matter. These create the spiral pattern, which slowly revolves around the centre of the Milky Way. The density waves perturb the interstellar gas, producing relatively dense clouds from which new stars are born, rejuvenating the Galaxy. As a result, massive, bright blue stars decorate the arms, but they are too short-lived to travel far from their birthplaces before they blow up and spew out cosmic rays.

Small stars like the Sun live long enough to orbit around the centre of the Galaxy many times. But as they don't cruise at the same rate as the spiral arms revolve, they repeatedly run into spiral arms and then re-emerge on the other side. The peak in the exposure to cosmic rays, occurring as the Sun and its attendant planets exit from the spiral arm, is due to many large stars manufactured at the leading edge of the spiral, and journeying a little way ahead of it before blowing up. Nir Shaviv estimated the climatic effects to be very large.

> The variations in the cosmic-ray flux arising from our galactic journey are ten times larger than the variations due to solar activity, at the high cosmic-ray energies responsible for ionizing the lower atmosphere. If the Sun is responsible for variations in the global temperature of about 1 degree Celsius, the effect of the spiral-arm passages should be about 10 degrees. That is more than enough to change the Earth from a hothouse where temperate climates extend to the polar regions, to an ice-house with ice-caps on the poles, as is the case today. In fact, the spiral-arm effect is expected to be the most dominant driver of climate changes over periods of hundreds of millions of years.

Four major arms, or segments of arms, cross the path of the Sun and Earth as they tour the Galaxy. Their names come from the constellations beyond which the different arms appear most conspicuously in the night sky. The minor corridor of bright stars called the Orion Arm, where we are now, protrudes from the major Perseus Arm towards which we are now heading, for an encounter 50–100 million years from now. Far in the future, the Earth will also revisit the Norma, Scutum-Crux and Sagittarius-Carina Arms.

Although astrophysicists can agree on the speed of the Sun in its galactic orbit, the rate of rotation of the pattern of spiral-arm pressure waves has been a matter of contention. Estimates over the past 40 years have ranged from half the Sun's rate of progress to a little faster than the Sun. To link encounters with spiral arms to climate changes, what matters is the relative speed of the Sun and the spiral arms. It fixes how frequently and when the peaks and troughs in cosmic rays occurred.

Can we reach so far into the abyss of time and space, to find out how the cosmic rays varied in the Sun's vicinity, hundreds of millions of years ago? The remarkable answer from Shaviv was: yes, we can. He found the rhythm of the cosmic rays by re-analysing German data on radioactivity in iron meteorites.

When asteroids collide far away in the Solar System, the fragments released into space can include lumps of iron. They continue to orbit around the Sun for hundreds of millions of years, and while they do so the impacts of cosmic rays manufacture radioactive atoms. Eventually a few of the fragments fall onto the Earth as iron meteorites. You might then try to gauge how long each meteorite has spent wandering in space by the amount of radioactive potassium atoms it contains, in proportion to stable atoms, but variations in the intensity of cosmic rays experienced in the Solar System falsify the results.

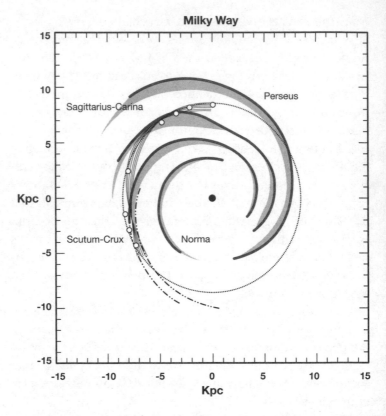

Milky Way

12. *The path of the Sun through the spiral arms of the Milky Way exposes the Earth to varying intensities of cosmic rays and alternations between hothouse and icehouse conditions. The climate record may help to reduce the uncertainties about the precise track of the Sun and the positions of the spiral arms, indicated here by multiple tracks and shading. (Scale in kiloparsecs, where 15 kpc means 49,000 light-years)*

The apparent ages of the iron meteorites bunch unnaturally at times when the cosmic clock ran slow, because cosmic rays were scarce. To rule out the possibility that bunching might be due in part to meteorites that originated in the same asteroidal event, Shaviv excluded cases where their character and ages were too similar. That still left him with 50 iron meteorites ranging up to a billion years in age, from which he could deduce that the cosmic-ray intensity went up and down in a cycle lasting 143 million years – give or take 10 million years – as the Solar System repeatedly passed through spiral arms of the Galaxy.

This result fitted uncannily well with the long-term record of climate change. Over the past half-century, geologists have recognised a slow alternation between hothouse and icehouse climates, and have gradually refined the dates. When Shaviv looked for a possible regularity in the climate data, the best fit came with a cycle of 145 million years – close to the length of the cosmic-ray cycle that he inferred from the meteorites.

Shaviv's analysis extended, as mentioned, over the past billion years. The first part of that period involves cosmic and climatic shocks of other kinds that are best left to Chapter 7. For the time being, let's see what astronomy can say about the experiences of the animals which first became well preserved as fossils of many different kinds in the Cambrian Period that began 542 million years ago. The whole interval since then is called the Phanerozoic Eon, meaning the time of conspicuous life.

Stresses of life in a spiral galaxy

At the beginning of the Cambrian Period the Sun and the Earth had just escaped from very icy conditions, after passing through the Sagittarius-Carina Arm of the Milky Way. Harsh climates can provoke evolutionary innovations for the sake of survival. The novel body plans of the animals originated among earlier generations of worms that had burrowed in the sea-bed, as James Valentine of the University of California in Berkeley first pointed out in the

1970s. The worms were relatively immune to seasonal and long-term climatic changes that left other animals hungry.

When a hothouse phase began in the Cambrian Period, ancestors of all the main branches of the animal kingdom were in place. With the Sun and Earth en route between two spiral arms of the Galaxy, cosmic rays were low and sea levels were high. Life flourished on the continental shelves. Among the great diversity of invertebrate animals were tadpole-like larvae that became precociously reproductive. They founded the dynasty that led to fishes and all the other animals with backbones.

The warm conditions persisted into the Ordovician Period, but were punctuated by a visit to the Perseus Arm of the Milky Way. About 445 million years ago the Ordovician ended with a sharp icehouse phase, when the sea level fell. Although relatively short-lived, the glaciers came right on cue, in Nir Shaviv's scheme, just as the Solar System emerged from the Perseus Arm and cosmic-ray intensities peaked.

Newcomers in the Silurian Period, in the warm aftermath of that stressful interlude, included the first plants and animals to live on land. The bony fishes appeared which would become the most successful of all animals with backbones. The hothouse theme continued into the Devonian Period that followed.

Uncertainty about the position of the next spiral arm to be visited, the Norma Arm, left a mismatch in Shaviv's original analysis between astronomical expectations and the cosmic-ray influx as gauged from the iron meteorites. A better interpretation of the spiral-arm pattern later removed most of the discrepancy. In any case, the meteorite data on cosmic rays agree well with geological evidence for a maximum chill occurring around 300 million years ago, at the end of the Carboniferous Period.

That icehouse episode, long known to geologists as the Permo-Carboniferous Ice Age, was not brief. It straddled the Carboniferous Period, which acquired its name from the large deposits of coal laid down in swampy forests. During that time the first reptiles appeared, as animals with backbones capable of living entirely ashore. But while trees flourished, widespread ice sheets and glaciers covered the continents that were then lying towards the South Pole. The icehouse continued into the early part of the Permian Period.

The latter part of the Permian and the whole Triassic Period that followed were a hothouse interlude when the Solar System was in a dark space between the Galaxy's arms. Catastrophe struck at the end of the Permian Period, 245 million years ago, in a mass extinction of species, presumably because of a collision of the Earth with a jay-walking comet or asteroid. It ushered in the Mesozoic Era, most famous for its dinosaurs. But the hothouse continued unabated, which shows in this case a de-coupling between climate and evolutionary change.

A passage through the Scutum-Crux Arm brought back cooler conditions during the Jurassic and early Cretaceous Periods. Among the novelties appearing then were the first flowering plants and the first birds. Starting about 120 million years ago, the Late Cretaceous hothouse followed – which leads us back to where we began the story, with the chalk of Møns Klint.

Anyone who rejoices in the re-convergence of the sciences, after a century of narrow-minded specialisation, may take satisfaction in marriages between the names of spiral arms of the Milky Way Galaxy and of the chilly geological periods associated with our planet's visits to them.

   Perseus Arm and Ordovician to Silurian Periods
   Norma Arm and Carboniferous Period
   Scutum-Crux Arm and Jurassic to Early Cretaceous
      Periods
   Sagittarius-Carina Arm and Miocene Epoch
   leading almost immediately (in geological terms) to
   Orion Arm and Pliocene to Pleistocene Epochs

What happened during that last transition, and the evolutionary consequences, will be explored in Chapter 7.
Meanwhile, a very striking example of an astronomical
prediction being quickly supported by geological and fossil
evidence comes from the origin of birds.

Feathers from a cold spell

When the first small dinosaurs and mammals made their
debuts around 230 million years ago the Sun and Earth
were roughly where they are now. During that time
the Solar System has gone full circle around the centre
of the Milky Way. For most of the journey the dinosaurs
were lords of the Earth and kept the mammals lying low,
although they themselves failed to complete the grand
tour.

   Conditions were warm in the Triassic Period when the
trip began. The dinosaurs' guide to the Galaxy, based on the
expected cosmic rays, would have shown the Scutum-Crux
Arm looming, and promising an icehouse episode in the
Jurassic and Cretaceous Periods. Yet generations of students
and fans of the giant reptiles have been told that living conditions on land throughout the dinosaurs' Mesozoic Era
were warm and free of ice. If any part of the geological
record was likely to confound the cosmic-ray reckonings,
this was it, as Nir Shaviv well knew.

When I first worked on the idea, I looked for glacial data and found a summary in a book from the 1970s that didn't include the mid-Mesozoic glaciations. I thought to myself, 'Oh well, cosmic rays don't explain all the climate variations.' Only later did I find another review of glaciations where it said that the mid-Mesozoic was colder than the epochs before and after. I had a grin on my face for a whole day since I'd found the lost icehouse epoch. At that point I knew that the theory had to be correct.

When Nir Shaviv first went public with his story about the spiral arms, in 2002, the clearest signs of mid-Mesozoic coldness came from ice-rafted debris on the ocean floor. Results assembled in 1988 by Larry Frakes of the University of Adelaide showed that floating ice released its grit in sub-polar regions. But the lack of any sign of ice on land left the mid-Mesozoic icehouse the least convincing of them all.

Early in 2003, just a few weeks after Shaviv's paper came out, an announcement came from Adelaide of the discovery of the first known ice on land from the Cretaceous Period. Neville Alley and Larry Frakes reported clay, small boulders and quartz grains crushed by a glacier near the Flinders Range of Western Australia. It dated from the Early Cretaceous, around 140 million years ago. So the dinosaurs really were on a climatic roller-coaster, as predicted correctly by the astronomy.

If you prefer living evidence of past chilliness, look at the birds. They are the only survivors of the dinosaur lineage, and they owe their existence to the mid-Mesozoic icehouse. By taking refuge in trees or marshland, small dinosaurs could escape the jaws of their fearsome cousins, but there was a snag. They lost their body heat far more rapidly than big animals did. The diminutive mammals had fur coats, but small dinosaurs in the cool Early Cretaceous kept warm by converting their scaly skin into down and feathers.

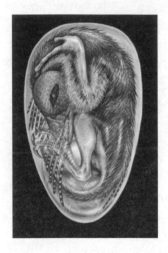

13. Out of the icehouse: found in China, a fossil bird chick in its egg 121 million years ago, and showing feathers, as clarified here by an artist. A cold climate due to a visit to one of the Galaxy's spiral arms encouraged small dinosaurs to acquire feathery down for warmth. Finding other uses for the feathers, some of them evolved into birds. (Zhongda Zhang, Institute of Vertebrate Paleontology and Paleoanthropology, Beijing)

At the same time as the Australian discovery of the Cretaceous glacier came conclusive proof from China for the existence of small feathered dinosaurs somewhat later in the Early Cretaceous, and of the evolution of some of them into birds with modern features. Their traces are preserved in the beds of former lakes in Liaoning Province in north-east China. The man who found the first fossil birds in the region, Zhou Zhonghe of the Institute of Vertebrate Palaeontology and Palaeoanthropology in Beijing, was clear about the implications.

> New discoveries produced interesting results that suggest feathers may not be unique to birds and that flight might have evolved from gliding by tree-dwelling creatures.

When you watch the birds that animate your gardens and skies, be glad that feathered creatures had plenty of time to establish themselves as a new class of vertebrates before the shocking events that wiped out all of the dinosaurs, big and small, 65 million years ago. Many birds as well as mammals survived the mass extinction when a comet or asteroid hit Mexico and vast amounts of volcanic lava gushed from India on the other side of the world. The icy episode about 75 million years earlier had encouraged small dinosaurs to experiment with feathery jackets – and to find out what else they could do with them.

The first evidence for the impact that killed the dinosaurs came to light in 1980. It took the form of rare elements of extraterrestrial origin detected in a layer of red clay that cuts across the limestone strata in a gorge near Gubbio in Italy. It was only one of a number of such events that re-directed the course of evolution, independently of long-term variations in the global climate. After a spell of chaotic weather following each impact, the climate reverted to the icehouse or hothouse conditions that prevailed before the comet or asteroid struck. At Gubbio, the limestone that walls the gorge below the clay continues to soar above it with only a slight change of colour – as if nothing much had happened.

A spat about carbon dioxide

Discover magazine counted Nir Shaviv's work on the Galaxy and climate as one of the top 100 science discoveries of 2003. The idea was very acceptable as a venture into uncharted territory where he didn't seem to be treading on anyone's toes. Shaviv himself had no idea that his cosmic-ray story would have repercussions on the debate about current climate change, until he teamed up with a prominent geologist, Ján Veizer.

Based at the University of Ottawa but working also at

the Ruhr University in Germany, Veizer had amassed a store of data on the abundance of heavy oxygen atoms in fossil shells of creatures living in the tropical oceans over the past 550 million years. These showed the tropical sea-surface rising and falling by about 4 degrees Celsius, more or less in step with the alternations between hothouse and icehouse conditions.

In 2000, with colleagues from Liège, Veizer reasoned that his data contradicted the widely-held belief that changes in the amount of carbon dioxide in the atmosphere were responsible for the temperature variations. Especially around 150 and 450 million years ago, which were icehouse times, high carbon dioxide concentrations would predict sea temperatures far warmer than those shown by Veizer's shell collection. Instead, the history showed a pronounced cycle of about 135 million years – similar to the cycle of around 143 million years that Shaviv expected from the crossings of the spiral arms. Shaviv included Veizer's graph when he published an extended version of his spiral-arm analysis in 2003.

The astrophysicist and geologist then realised that by getting together they could make more precise estimates of effectiveness of cosmic rays in climate change. They collaborated on a provocative article entitled 'Celestial driver of Phanerozoic climate?' published by the Geological Society of America in *GSA Today*, which is widely read by geologists. Besides pooling their own data, they explained the Danish results on cosmic rays and clouds – perhaps, for many readers, the first time they had heard of them.

Shaviv and Veizer concluded that the link between the Phanerozoic climate and cosmic rays was inescapable, while the effect of carbon dioxide on climate had to be less than generally claimed. From the mismatches in the geological record between carbon dioxide levels and sea temperatures, they judged that a future change in temperature, due to a doubling of carbon dioxide, would be lower than predicted by the Intergovernmental Panel on Climate Change. Overnight, Shaviv and Veizer found themselves rated *persona non grata*.

Six months later, a posse of eleven scientists assailed their heresy in the geophysical magazine *Eos*. The lead author was Stefan Rahmstorf of the Potsdam Institute for Climate Impact Research. The article began by doubting the effect of cosmic rays on climate, relying on an out-of-date attempt to deny it. And as the critics had not read their own article very carefully, Shaviv and Veizer were able to rebut many other points by repeating what they wrote originally.

The debate was too intricate and arcane to summarise here, but one example will give its flavour. Rahmstorf and his fellow critics suggested that a graph of sea-surface temperatures had been manipulated to emphasise variations matching the cosmic rays. Here the rebuttal was like a rap on the critics' fingers: 'The calculated temperature trends ... were already published in Veizer et al. (1999, 2000), in total ignorance of Shaviv's future work.'

GSA Today carried a better reasoned commentary, entitled 'CO_2 as a primary driver of Phanerozoic climate'. The five authors were led by Dana Royer of Penn State University. They said that the temperatures deduced from heavy oxygen atoms in old carbonate deposits ought to be corrected for the degree of acidity prevailing in sea water at the time. Then, they suggested, the match between temperatures and carbon dioxide became much better. 'Changes in cosmic ray flux may affect climate but they are not the dominant climate driver on a multi-million-year time scale.'

You can go to the root of the issue very easily. The

carbon dioxide levels show just two bumps and two dips over 550 million years, while the cosmic-ray levels show four of each. As there were four hothouse and four icehouse intervals, the pattern supports Shaviv and Veizer's identification of the cosmic rays as the primary driver. But the differences in severity of the icehouses suggest that something else was going on.

An attempt to end the spat about which was more important – cosmic rays or carbon dioxide – came from Klaus Wallmann of the Geomar research centre in Kiel. Writing in the journal *Geochemistry Geophysics Geosystems*, he declared that he couldn't reproduce the temperature trends (corrected for acidity) in his calculations without adding the cooling effect of cosmic rays. On the other hand, carbon dioxide had a notable role, Wallman said, in intensifying or reducing the climate changes.

> Warm periods (Cambrian, Devonian, Triassic, Cretaceous) are characterised by low cosmic-ray levels. Cold periods during the late Carboniferous to early Permian and the late Cenozoic [i.e. present times] are marked by high cosmic-ray fluxes and low carbon-dioxide values. … The two moderately cool periods during the Ordovician–Silurian and Jurassic–early Cretaceous are characterised by both high carbon-dioxide and cosmic-ray levels so that greenhouse warming compensated for the cooling effect of low-altitude clouds.

How strong was the effect of carbon dioxide in the remote past? In the dips, 300 million years ago and in the present icehouse, the amount of carbon dioxide in the air has been just a few hundred parts per million, but in the bumps it was around 5,000 and 2,000 parts per million. For a translation into the terms used in reckoning present-day climate change, you have to ask, what would be the rise in temperature for an increase from 280 to 560 parts per million – a doubling compared with pre-industrial levels? The Intergovernmental Panel on Climate Change thought that this climate sensitivity was probably in the range 1.5 to 4.5 degrees Celsius.

Shaviv and Veizer's initial suggestion from their 500-million-year study was that the climate sensitivity could be as low as 0.5 degrees. But they acknowledged that corrections for acidity were needed, although they thought that Dana Royer and his colleagues were overstating them. Shaviv and Veizer also stressed that the count of heavy oxygen atoms used to reckon the temperature had to be corrected for the amount of ice in the world, because making ice sheets increases the heavy oxygen remaining in the sea. They then came up with a revised estimate for the climate sensitivity of about 1.1 degrees Celsius.

This is in line with an assessment of the current atmosphere by an eminent meteorologist, Richard Lindzen of the Massachusetts Institute of Technology. He has long favoured a moderate value for the climate sensitivity, as he explained in testimony to the British House of Lords in 2005.

> If the major greenhouse substances – water vapour and clouds – remain fixed, a doubling of CO_2 should lead on the basis of straightforward physics to a globally averaged warming of about 1 degree Celsius.

So the ancient fossils guided Shaviv and Veizer to a figure very close to that cited by Lindzen on the basis of modern studies. That is very thought-provoking. Svensmark hesitates to put a number on carbon dioxide's climatic potency, and wonders whether it remains the same on all geological timescales and for a huge range of variations in carbon dioxide concentrations. Be that as it may, Shaviv and Veizer's result is well below the climate sensitivity needed to justify the more alarming predictions of man-made global warming in the 21st century. It therefore matches Svensmark's general optimism about the implications of the cosmic-ray story for the fate of the planet in the industrial era.

Using sea-shells as telescopes

When Nir Shaviv's enthralling account of the part played by cosmic rays in the history of the Earth's climate came out in 2002, Svensmark pondered the implications and started drafting related papers of his own. But the poor quality of the geological records that he had to hand impeded him until, at a meeting in Hawaii in 2005, Shaviv pointed him to a better database. The experiment in the basement had been a big distraction too. But when the first complete sets of results from SKY were available and interpreted, Svensmark was able to attend more closely to the revelation that stars and rocks have followed the same song-sheet since time immemorial.

He was struck by the contradictory opinions among astronomers about the Milky Way and the timing of the Sun's encounters with its spiral arms. These were in some ways more discouraging than any geological uncertainties about the changing climate. He decided to turn the reasoning around and to use Ján Veizer's fossil record of sea temperatures to improve the astronomy: 'For fun I could call it, "How to measure the mass of the Galaxy with a thermometer."'

The shelly inhabitants of the sea act as natural sensors that record the ever-changing starry environment, and they did so long before any man-made astronomical instruments existed. Interpreted with hindsight, they were telescopes that registered the intensities of cosmic rays by their uptake of heavy oxygen when they were alive. So the idea of using sea-shells for astronomical purposes is not fanciful.

The fossils show relatively small variations in climate with a more rapid rhythm than those due to the visits to the spiral arms. The reason is that the Sun behaves like a playful dolphin. While it orbits in the Milky Way, it also rises and plunges and rises again repeatedly, through the disc of stars that surrounds the central bulge of the Galaxy. The mid-plane that defines the disc is not just a mathematical fiction. Cosmic rays concentrate there, because the magnetic field that guides them is held in place by the gravity that keeps stars and gas clouds confined close to the disc.

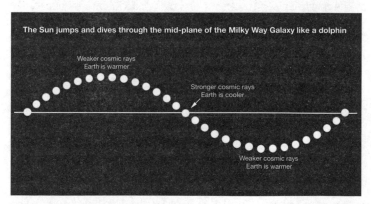

The Sun jumps and dives through the mid-plane of the Milky Way Galaxy like a dolphin

Weaker cosmic rays
Earth is warmer

Stronger cosmic rays
Earth is cooler

Weaker cosmic rays
Earth is warmer

14. *Past wobbles in the climate linked to the motions of the Sun can help to improve astronomical knowledge about the Galaxy.*

So the cosmic rays are more intense on the Earth whenever the Sun crosses the mid-plane, whether going up or down, which happens at intervals of about 34 million years. When it leaves the mid-plane, it climbs about 300 light-years away from it, curves over and begins to fall back. At that phase the cosmic rays are weaker. These variations occur regardless of whether the Solar System is inside or outside a spiral arm, but the pace of the diving should be quicker inside an arm, because of the stronger gravity of concentrated gas. The sea temperatures set tight limits on the timing, because the coolest phase of each 34-million-year cycle, well dated by geologists, corresponds with a crossing of the mid-plane.

Svensmark did not pre-judge the question of how quickly the Sun moves in relation to the pattern of spiral arms. That was one of the things he wanted to measure, by taking Veizer's temperature record for the past 200 million years and seeing how best to relate the Galaxy to it. The mathematical procedure was like having a ready-made suit and searching for the person who would fit in it best. Only one combination of key numbers describing the galactic environment gives the correct dolphin-like motions of the Sun.

A string of answers describing the nearer parts of the Milky Way and the Sun's passage through it came from Svensmark's analysis. The relative speed of the Sun and the rotating pattern of spiral arms is 12 kilometres per second. The visit to the Scutum-Crux Arm occurred 142 million years ago, and that to the Sagittarius-Carina Arm, 34 million years ago. The arms were about 1,170 and 910 light-years wide, respectively. The density of matter in the spiral arms is 80 per cent higher than in the inter-arm region – and this is indeed using temperature changes to weigh the Galaxy!

None of the numbers falls outside the ranges of previous suggestions. But where there was much uncertainty before, the fossils tell the astronomers which numbers are right.

This successful inversion of the reasoning, from climate to astronomy, confirms that the Earth's climate is firmly under the control of a variable galactic thermostat. The next chapter will tell how the starry sky had an even more emphatic impact, revealed in the unexpected convergence of two other big discoveries of recent decades.

6 Starbursts, tropical ice and life's changing fortunes

Episodes when the Earth froze over amaze geologists · They occurred when, and only when, star-making peaked · The Galaxy's baby booms intensified the cosmic rays · Early protection by the young Sun helped life to begin · In icy periods, life lurches between glut and dearth

For those who dream of finding life beyond the Earth, Mars has always been the prime hunting ground. But nowadays one of the moons of the giant planet Jupiter pricks their curiosity. Although entirely covered by ice, Europa supposedly conceals a liquid ocean. Space explorers would dearly like to drill through the ice and look for possible signs of living organisms. If that seems far-fetched, consider that on more than one occasion in the distant past the Earth looked rather like Europa. An interplanetary visitor who drilled through the ice at such a time to see if anyone was at home would have found the planet's microbial crew cowering to survive the rigours of Snowball Earth.

That such extreme conditions could have occurred on our home planet was first mooted in the 1960s. Ice sheets and glaciers are normally confined – except on high mountains – to regions not very far from the poles. In the most severe periods of the most recent ice ages, ice sheets penetrated little closer to the Equator than Manhattan Island in New York. But Brian Harland of Cambridge University noticed that in deposits dating from around 600 million years ago the evidence of glaciation was extraordinarily widespread, as if the whole world was icy then.

Might the ever-wandering continents have clustered near the poles, where ice is unsurprising? That possibility could be tested, and ruled out. Evidence of where the continents lay in the past comes from the magnetic memories of rocks. If they formed near the poles, the magnetism of the rocks slants downwards, while near the Equator the magnetism will be roughly horizontal. Rock magnetism suggested that, so far from being grouped near the poles, the continents were crowded in the tropics.

In 1986, George Williams and Brian Embleton in Australia used the magnetism in grains of iron oxide dropped from ancient ice to show that they were released within a few degrees of the Equator. A few years later, Joseph Kirschvink of the California Institute of Technology confirmed this result in magnetism associated with other rock formations in Australia produced by ice action, and well dated as 700 million years old. He called it 'bullet-proof evidence'.

It now seems clear that these extensive, sea-level deposits ... were formed by widespread continental glaciers which were within a few degrees of the equator. The data are difficult to interpret in any fashion other than that of a widespread, equatorial glaciation.

Kirschvink invented the name Snowball Earth for that dire climatic state. You have to visualise ice sheets, glaciers and frozen seas even at the Equator itself. The degree of ocean freezing is still debated. Some investigators imagine vistas of ice a kilometre thick or more, others prefer a 'slushball' picture with drifting sea ice and icebergs. Either way the impact on life was severe.

15. *This moon of the giant planet Jupiter, called Europa, is covered with cracked ice sheets. Our own planet may have looked much the same during its coldest-ever episodes, called Snowball Earth. (NASA/Voyager 2)*

Evidence from all the world's continents unpacks into about three separate snowball episodes in the interval 750 to 580 million years ago. Worms that survived by scavenging the sea-bed detritus evolved the body-plans that made possible the explosion of animal life mentioned in the previous chapter, when the world became reliably warmer again in the Cambrian Period that started 542 million years ago.

Those cold Neo-Proterozoic times, as geologists call them, were not the only occasion of such radical events involving ice and evolution. By the end of the 20th century, geologists had amassed evidence from South Africa, Canada and Finland that confirmed two Snowball Earth episodes between 2,400 and 2,200 million years ago, in Palaeo-Proterozoic times. Our planet was then only half its present age.

Remarkable souvenirs from around that time include the world's largest deposits of iron and manganese ore, produced by the action of oxygen on those metals dissolved in sea water. The whole planet rusted. Many ancient lineages of bacteria were wiped out on Snowball Earth, but novel microbes called eukaryotes survived the massacre.

These were single-celled fungi, algae and animal-like grazers distinguished by their use of cell nuclei to encapsulate their genes. By 1,800 million years ago some eukaryotes had taken in oxygen-handling bacteria to serve as power-stations, in cells of the modern kind now found in every plant and animal. The descendants of those bacterial lodgers are present in your body as mitochondria. Their antiquity is manifest. They originated before sex was invented, and you inherited them only from your mother.

The big geochemical and biological events associated with the climatic extremes of Snowball Earth prompt debates about cause and effect. One scenario for the Palaeo-

Proterozoic freezes was that over-zealous production of oxygen by bacteria caused the rusting episode and somehow triggered the icing of the planet by altering the composition of the atmosphere.

But the main challenge for anyone wanting to account for the snowball events is to say why they occurred at particular times in the long history of the Earth, in two relatively brief windows of time around 2,300 million and 700 million years ago. A complete solution to the conundrum should also say why, between those occurrences, our planet was completely free from ice for a billion years.

The chilling stars provide the only explanation for Snowball Earth that specifies these timings. After Nir Shaviv in Jerusalem had accounted for the hothouse–icehouse alternations in climate in the past 500,000 years, by visits to the spiral arms of the Milky Way, his next step was to link the snowball events to episodes of star-making in the Galaxy. These boosted the cosmic rays to such exceptional levels that the Earth became cloudy and sunless enough to freeze over.

Baby booms among the stars

While the evidence for Snowball Earth was astonishing the geologists, the astronomers were taken aback by galaxies far warmer than expected. First detected by the Dutch–US–UK Infrared Astronomy Satellite in 1983, they give off intense, though invisible, rays. By 1998, when Europe's Infrared Space Observatory had examined hundreds of these ultra-luminous objects in detail, Reinhard Genzel of the Max-Planck-Institut für Extraterrestrische Physik at Garching could announce the astronomers' conclusion.

This is the first time we can prove that most if not all of

the luminosity of the ultra-luminous infrared galaxies comes from star formation. To understand how, and for how long, such vigorous star formation can occur in these galaxies is now one of the most interesting questions in astrophysics.

The vehicles of this frenzied activity are now called starburst galaxies. The infrared rays come from warm dust produced by the explosions of large numbers of massive, short-lived stars. In most if not all cases, the starbursts are due to collisions between galaxies. Pictures from the big telescopes show many such stupendous traffic accidents.

Even with many billions of stars involved, in each of the participating galaxies, the spaces between stars are so wide that the chances of a direct collision of two stars are slim. Instead, the high-speed encounter of gas carried by the galaxies creates shock waves that compress the gas and provoke its collapse into new-born stars. The milder, long-playing perturbations that create the bright spiral arms of our Galaxy produce about two new stars a year. In a starburst the birth-rate can be 50 or a hundred times higher.

Most galaxies waltz together in large clusters. We see only a snapshot of the cosmic ballroom, because the steps in the dance can take hundreds of millions of years. The choreographer is gravity – not just the mutual tugs of the masses of stars and black holes that make up the galaxies, but the much stronger gravity of mysterious dark matter that binds the clusters together. Besides the currently active starburst galaxies, most other large galaxies have experienced such events in the past. In some cases they have exhausted so much of the interstellar gas that star-making has ceased.

Collisions are inevitable in the densely packed clusters

of galaxies. With many short-lived stars ending their lives in supernova events, the intensity of cosmic rays within a starburst galaxy is so high that you have to wonder whether life could survive on the surfaces of its planets. What makes matters worse is that a large cluster of galaxies traps almost all the cosmic rays ever made within its member galaxies, instead of letting them leak away into the Universe at large.

Perhaps life can survive only in thinly populated parts of the cosmos, where it is relatively immune from large starbursts and long-playing cosmic rays. Our own galactic home, the Milky Way, is lucky that way. Although telescopes reveal more than 30 galaxies within about 5 million light-years, in an assembly called the Local Group, most of them are very small.

Only three nearby galaxies are visible to the naked eye. The Large and Small Magellanic Clouds are small, as galaxies go, but close enough in the southern sky to have been very plain to the explorer Ferdinand Magellan in 1519. They look like untidy scraps of the Milky Way. In the north, the rather faint Andromeda Galaxy was noted by Persian astronomers in the 10th century. Now we know that it's a big spiral, the Milky Way's sister, almost 3 million light-years away. The Andromeda Galaxy is heading in our direction. Perhaps it will collide with the Milky Way and eventually merge with it in colossal starburst events – but that would be 5 billion years from now.

You don't need an outright collision to provoke star-making. When two galaxies pass close to one another, gravity stirs the pot in both of them by raising tides and pressure waves. Several of the small galaxies in the Local Group are satellites of the big ones, and the Magellanic Clouds orbit around the Milky Way. They are the most

16. *Traffic accidents are common among galaxies, and here two vast assemblies of stars have collided in the constellation Corvus, to make the pair called the Antennae. Infrared rays pouring from them tell of great starbursts of stellar birth and death provoked by the collision. Intense cosmic rays are a by-product, so it is fortunate that nothing quite so spectacular has happened to the Milky Way since life began. (François Schweizer, CIW/DTM)*

likely candidates for close encounters capable of causing starbursts. Although not on the scale of the events in the ultra-luminous infrared galaxies, starbursts in our Galaxy can raise the rate of star-birth and star-death sufficiently to cause a marked increase in the intensity of cosmic rays.

To find out when the star formation rate has changed, astronomers take a census of the stars. If you find in a human population an unusually large number of people in a certain age group, you'll know that they were born during a baby boom. So it is with stars. But to calculate the age of any star, astronomers have first to measure how far away it is. The distances of many stars became much better known in 1997, with the release of results from Europe's star-mapping satellite Hipparcos.

Astronomers in Brazil and Finland used the Hipparcos data to help them to compare the ages of some 500 nearby stars. By 2000, Helio Rocha-Pinto and his colleagues were able to report clumps in the ages that told of several stellar baby booms during the Galaxy's long history. The survivors seen today are necessarily modest, long-lived stars, but their massive cousins would have soon exploded and generated cosmic rays in abundance during those periods of high rates of star formation.

One of the baby booms fell in the period 2,400–2,000 million years ago. An unusual number of stars of the same age in the Small Magellanic Cloud provides supporting evidence and points the finger at a neighbouring galaxy that may have come close enough to provoke the action in the Milky Way. On the other hand, some astronomers suspect that the Large Magellanic Cloud was the perpetrator. Knowledge of the comings and goings of the Magellanic Clouds and other small neighbours in their clumsy dance is sketchy at best. So is the timetable of their close approaches or 'perigalacticons'. Timings will remain uncertain until better measurements of the present motions of the nearest galaxies become available from Europe's next star-mapping spacecraft, Gaia, by about 2015.

Meanwhile, what stands out is the correspondence in time between the early Snowball Earth episodes of about 2,300 million years ago and Rocha-Pinto's starburst in the period 2,400 to 2,000 million years ago. There are reasons for suspecting that the two events were connected, by the unusually high cosmic rays to which the Earth was subjected. But if this was more than a chance coincidence, then the ice-free interval that followed should be associated with a scarcity of stars born at that time. For Shaviv this was a key point in his argument.

> The long period of 1 to 2 billion years before present, during which no glaciations are known to have occurred, coincides with a significant paucity in the past star formation rate.

And the later Snowball Earth episodes starting around 750 million years ago should also be linked to another stellar baby boom. Rocha-Pinto's census of Hipparcos stars does indeed show star-birth much reduced between 2,000 and 1,000 million years ago. But the rate of star formation that follows the lull, in this census, is not very impressive. More persuasive are the results of another survey announced in 2004 by Raúl de la Fuente Marcos of Suffolk University, Madrid, and Carlos de la Fuente Marcos of Universidad Complutense de Madrid. They used data on groups of stars called open clusters, as catalogued by astronomers over many years, to infer among their other results a starburst around 750 million years ago. The Fuente Marcos pair noted its timeliness for Shaviv's story.

> The Snowball Earth scenario appears to be connected with the strongest episode of enhanced star formation recorded in the solar neighbourhood during the last 2,000 million years.

Here is extraordinary support for the idea that cosmic rays have controlled the climate throughout the Earth's history. When a hypothesis is false, new experiments and observations will tend to quarrel with it, but with a good theory the reverse is true. It looks better and better as the facts become more exactly known.

The paradox of the faint young Sun

The deep-freeze of the Snowball Earth events would have been even more severe if the Sun's magnetic shield against cosmic rays had not been stronger in the remote past. The influx of ground-penetrating cosmic rays 750 million years ago was a few per cent down on what it would be today, if the same starburst repeated itself, because the solar wind was stronger then. And 2,400 million years ago the Sun's shield was strong enough to cut the influx by 20 per cent.

Going even further back in time, the Sun was very different from the way it is now. Astronomers know this from studying young sun-like stars, as well as by theories of the Sun's internal history. When first born from its own dusty gas cloud about 4,600 million years ago, together with its family of planets, the Sun rotated at a rate at least ten times faster than today. Its magnetic activity was very vigorous and the solar wind was denser. As a result, virtually no cosmic rays could reach the vicinity of the new-born Earth.

That was just as well, climate-wise, because the young Sun was cooler and giving off much less sunlight than it does today. It grew brighter only gradually, over billions of years, as the nuclear reactions in its hot centre filled an expanding core with manufactured helium. Early in its life, the Sun radiated only 70 per cent of its present sunlight. The Earth's surface rocks were probably molten at the time of its formation, but as soon as it was cool enough for liquid water to settle, the young planet might have frozen over because of the faintness of the Sun. That did not happen.

The early crust was almost completely destroyed and repeatedly recycled by a very heavy bombardment of impacting comets and asteroids – leftover raw material from the building of the planets. This hellish or Hadean Eon – so labelled by the geologists – lasted for 800 million years. From the very young Earth, just a few mineral grains survive, notably zircons found in Australia. The oldest, 4,400 million years old, is a fragment of zircon identified in 2001. Zircons are usually associated with granite, which requires liquid water for its formation, and a high proportion of heavy oxygen atoms in the zircon gives more direct evidence for a wet origin.

From the interval beginning 3,800 million years ago, the Archean Eon, many more rock formations survive, often showing clear signs of accumulation on ancient sea-beds. By that time sunlight had increased to about 75 per cent of its present strength, but that was still very faint by modern standards. If all other things were equal, the average surface temperature would have been, not 10 degrees Celsius as it is today, but *minus* 15 degrees. Even at the onset of the Proterozoic Eon, 2,500 million years ago, sunlight was still down at about 83 per cent, only promising global temperatures around *minus* 5 degrees. If geologists had come upon the evidence for Snowball Earth from 2,400 million years ago before they knew about older and warmer times, they might not have been so surprised. They could have simply blamed the faint Sun for the very icy conditions.

Instead they have a problem. Ever since 1972, when the American astronomer Carl Sagan and his colleague George Mullen first drew attention to 'the faint young Sun paradox', experts have tried to explain the warm conditions early in the Earth's history. Some have suggested that the Sun evolved differently from the sun-like stars, but increasing knowledge of the solar interior makes that idea impossible to cling to. Another proposal is that the atmosphere was dense and quite different from today's. Large amounts of water vapour, carbon dioxide, methane and/or other gases supposedly exerted a greenhouse effect sufficient to keep the world warm enough for liquid water. This conjecture has been repeated so often, over more than 30 years, that some experts treat the primordial greenhouse as if it were a fact.

What the make-up of the atmosphere was, when the Earth was very young, no one knows. All kinds of recipes are proposed and debated, with little constraint by evidence from the rocks or from other planets and moons in the Solar System. Even supposing you were granted a good snapshot of the atmosphere at one early moment, it would not be trustworthy, because the most violent impacts of the Hadean Eon probably blasted the pre-existing atmosphere away. Its replacements may have been quite different. By the time real data are available from the rocks, the very high concentrations of carbon dioxide in the air that some experts propose would have made the oceans acidic. Ján Veizer in Ottawa thinks that the evidence is against that.

Liquid water under a faint young Sun – that's all that is really clear. Only one escape from the paradox requires no ad hoc invention of special circumstances or special climatic mechanisms. From everything said so far about cosmic rays and clouds, it follows very simply that, with very few cosmic rays reaching the lower atmosphere because of the magnetic vigour of the young Sun, there would have been very few low-level clouds to cool the Earth.

This idea came to Nir Shaviv in Jerusalem, as he traced the climatic story of life in the Milky Way Galaxy backwards in time, through the icehouse episodes of the spiral-arm encounters and the Snowball Earth events due to high star-formation rates. He summed up the proposition at the end of 2003.

Standard solar models predict a solar luminosity that gradually increased by about 30 per cent over the past 4.5 billion years. Under the faint Sun, Earth should have been frozen solid for most of its existence. Yet, running water is observed to have been present since very early in Earth's history. This enigma ... can be partially resolved once we consider the cooling effect that cosmic rays are suspected to have on the global climate and by considering that the younger Sun must have had a stronger solar wind such that it was more effective at stopping cosmic rays from reaching Earth.

Svensmark had been thinking along similar lines some years earlier, but put the problem aside. Taking it up again, he estimated the benefits for the climate. At present, the low clouds reflect about 5 per cent of the incoming sunlight. If they were absent on the young Earth, the increase in sunlight reaching the surface would be equivalent to making the Sun a billion years older and brighter. The global mean temperature 3,800 million years ago would be raised from about *minus* 15 degrees Celsius to about *minus* 10 degrees, greatly reducing the contribution needed from greenhouse gases to ensure that the warmest parts of the world could support liquid water.

Glut and dearth reported by carbon atoms

Black specks of carbon found in rocks of Greenland 3,800 million years old are probably the oldest known traces of life on the Earth. They occur in the remains of a thick layer of clay. It was made by sediments that slowly piled up on the bed of a primordial sea, and is now exposed between ice sheet and ocean near Godthab on Greenland's western coast. Vast numbers of microscopic globules of graphite in the clay seem to be remnants of bacteria that thrived in the water when the world was young.

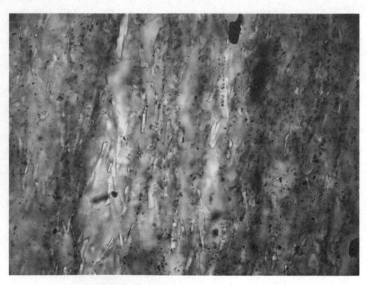

17. *The earliest known traces of life are minute carbonised globules in Greenland rocks 3,800 million years old, revealed as black spots by a microscope. A shortage of heavy carbon-13 atoms in the carbon indicates that they were bacteria that grew by taking in carbon dioxide and preferred the lighter kind of carbon. The existence of life in the sea so long ago tells of a friendly climate. Although the Sun was rather faint when the Earth was young, it kept the cosmic rays at bay. (Minik Rosing, Geological Museum, Copenhagen)*

For Minik Rosing, a Greenlander who directs the Geological Museum in Copenhagen, the globules showed the choosiness of living things about different atoms of carbon, the prime element of life. Nowadays when bacteria and algae of the sea-surface plankton grow by taking in carbon dioxide dissolved in the water, they prefer molecules containing the commonplace carbon-12 atoms. The heavier carbon-13, which crops up in one carbon dioxide molecule in 90, is likely to be rejected.

As a result, the living organisms contain less than the standard ration of carbon-13. It was just such a depletion of carbon-13 that enabled Rosing to suggest that his black globules came from material that was once alive.

> The precursor organic detritus of the graphite globules could have been derived more or less continuously from planktonic organisms that sedimented from the surface waters.

After announcing that discovery in 1999, Rosing continued his detective work on the conditions in the ancient sea where the early organisms lived. By 2004 he and his colleague Robert Frei had shown that the water apparently contained free oxygen. This time the clue came from traces of lead of various atomic weights, made by the decay of heavy radioactive elements. The proportions of the different lead atoms showed that 3,800 million years ago uranium had been present in the sea water, but not thorium. These elements remain firm companions except in the presence of oxygen, which makes the uranium soluble.

That implied that sophisticated bacteria were already alive. A billion years sooner than other experts expected, some of the bacteria were using the modern method of photosynthesis, whereby the energy of sunlight splits water molecules. The hydrogen goes into the carbon compounds needed to power and build living cells, while oxygen molecules escape into the environment.

By the time of the oldest surviving sedimentary rocks, life was no fugitive or paltry affair. Some experts have speculated that early life relied, not on sunlight, but on chemical energy from the Earth's interior, as seen today around volcanic vents on the ocean floor. But the picture from Greenland is of living things that comprised a fully-fledged, large-scale system worthy of the name biosphere, meaning the totality of life on and near the Earth's surface. Rosing was confident about the meaning, when he put his carbon-13 and uranium results together: 'What this demonstrates is that the Earth had a functioning biosphere before 3.7 billion years ago.'

Despite the faintness of the young Sun, its light was feeding large amounts of energy into the system. Sun-powered life has been a major participant in geological action on the planet ever since. In Rosing's opinion it's no coincidence that the first signs of continental granite are found in the same district of Greenland as the black globules denoting life.

Having worked our way back to its beginning, we can now trace life forward towards the present, and find another thought-provoking link between cosmic rays and the changing fortunes of life itself. In the course of its very long history, the biosphere has sometimes prospered and sometimes faltered, with huge swings between high productivity and dearth. For geological time-travellers the carbon-13 atoms, which were the giveaway for organic growth in the first place, can recapitulate the entire boom-or-bust story of life.

The Greenland globules showed the depletion of carbon-

13 characteristic of the microscopic sea-plants, bacteria or algae, which grow by taking in carbon dioxide from the surrounding water. When life is abundant, the water becomes noticeably enriched in the rejected carbon-13. Mementoes of the enrichment are preserved in limestone – in ancient carbonate rocks made from the carbon dioxide of the time. The fastidiousness about carbon atoms during photosynthesis does not apply in shell-making.

Whether produced from the carbonate shells of sea-creatures large and small, or purely by chemical action, the limestone indicates the overall activity of life in the sea, according to the proportion of carbon-13 remaining in the water. Geophysicists began routine measurements of carbon-13 half a century ago, as a by-product of their investigations of heavy oxygen, oxygen-18, in carbonate sediments. They knew that the oxygen-18 would help them to gauge temperatures in the past. It was easy to collect data on carbon-13 at the same time. To begin with, the investigators were unclear about what the results would tell them. They soon realised that carbon-13 gave them new insight into past changes in the overall state of life on the planet – in other words, variations in the productivity of the biosphere.

In terms of billions of tons of organic matter newly created each year, you might expect life to be at its most productive when the climate is warm and balmy. That is not the case. A peak in carbon-13 shows that the highest productivity in the past 500 million years occurred during the latter part of the Carboniferous Period, 300–320 million years ago. Then the cosmic rays were intense, as a result of the Earth's visit to the Norma Arm of the Milky Way. Vast ice sheets covered the southern continents.

Why should life have prospered then? Probably for the

same reasons as it flourishes in the present-day icehouse. The contrast in temperature between warm tropics and glaciated poles creates strong winds and vigorous ocean currents, and the effect on life of global weather patterns is visible from space. Satellites can gauge the productivity of the sea surface by the abundance of chlorophyll it contains. They observe vast areas of sub-tropical oceans that contain much less life per square kilometre than do the stormy mid-latitudes and sub-polar seas, where the surface waters are better re-supplied with essential nutrients such as phosphorus. In calmer periods of the Earth's history the scarcity of nutrients was more widespread, and it restricted life to a moderate abundance.

Not surprisingly, the episodes of Snowball Earth around 2,300 and 700 million years ago sometimes showed extremely low levels of carbon-13 in carbonates that formed when the global ice had virtually halted photosynthesis and dead organisms had returned their carbon-12 to the system. But these times of great dearth were interspersed with bursts of high productivity. Whenever there was any let-up in the extreme iciness, life in the sea bounced back with vigour. Besides the effect of released nutrients, as in the Carboniferous Period or nowadays, fertilisation by unusually high levels of carbon dioxide available for incorporation into organic material may have boosted growth in the Snowball Earth intermissions.

These hints from the drama of life as told by carbon atoms led Svensmark to a new overview of changing fortunes over billions of years. It turns out that the Earth's stellar environment in the Milky Way determines whether the lurches between scarcity and abundance in ocean productivity are modest or extravagant.

Life's variability and the cosmic rays

Apart from these logical changes in carbon-13 through geological time, between low values when life was nearly extinguished by Snowball Earth events and high values in productive intervals like the Carboniferous Period, something else has been going on. Non-stop fluctuations suggest an almost fickle relationship between geology, climate and biology. And the fluctuations themselves vary in intensity from one phase of the Earth's history to another.

In 2005 Svensmark noticed that the variability of carbon-13 was closely linked to variability in sea temperatures as gauged by oxygen-18. Big and frequent fluctuations in life's productivity accompanied frequent changes in the climate over the past 500 million years. But when Svensmark looked further back in time the variability of the biosphere was sometimes much greater.

He was astonished to see that the variability peaked 2,400 to 2,000 million years ago, around the time of the early Snowball Earth episode, and just when the cosmic rays were most intense as a result of the starburst event in the Milky Way. Svensmark pinned down the link over 3,600 million years, divided into 400-million-year segments, by comparing changes in carbon-13 variability with the calculated cosmic-ray intensities.

The match was almost unbelievably good, with a correlation of 92 per cent. 100 per cent would mean a perfect correlation. One reason for the similarity may be that when cosmic-ray intensities are very high, large variations in intensity also occur. That implies not just cold conditions when cosmic rays are high, but also bigger swings in climate between rather warmer and rather colder, as the Sun and Earth orbit around the Galaxy and the contrast between the spiral arms and the inter-arm gaps becomes greater.

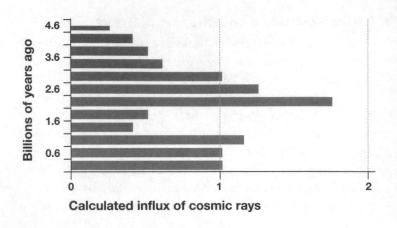

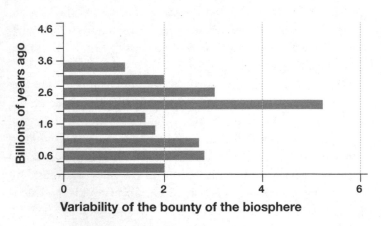

18. These look-alike graphs from completely different sources – astronomy and geology – revealed that life's variability depends on the Earth's galactic environment and the general intensity of cosmic rays. The upper one shows the influx of cosmic rays compared with the present rate, taken to be 1, while 'the variability of life's bounty' is a statistical gauge of the fluctuations in the proportion of carbon-13 atoms in marine sediments.

Around 3,400 million years ago the young Sun's action in warding off cosmic rays kept them at a low level, and the productivity signalled by carbon-13 varied relatively little. Between 3,200 and 2,800 million years ago, the star formation rate was similar to today's. So were the variations in biological productivity in the sea.

How curious! Only bacteria existed then, and now a fleet of much cleverer organisms is at work, supporting food chains all the way up to the top fishes and whales. Yet the overall responsiveness of the early bacteria and modern ecosystems to climate change remains much the same, as judged by the departures from the average rate of fixing carbon dioxide for growth.

About 2,800 million years ago the intensities of cosmic rays rose to a higher level, bringing greater variability in climate and in the productivity of the biosphere. At the peak of the starburst 2,400–2,000 million years ago, which brought the first Snowball Earth events, the cosmic rays were stronger still, and so were the carbon-13 variations that first caught Svensmark's eye.

From 2,000 to 1,200 million years ago, the cosmic-ray flux was again very low and the productivity of the biosphere varied very little. The build-up to the next stellar baby boom, starting 1,200 million years ago, re-awakened the variability, even as it ushered in the snowball events starting 750 million years ago. This was the time of the 'big bang' in evolution that invented multi-celled eukaryotes – ancestors of the animals and the higher plants. The variability of the biosphere was relatively high 800 million years ago. Since then it has declined, bringing it back to where it was 3,000 million years ago.

This novel history of the biosphere, told by interpreting the carbon-13 data alongside the astronomical stories, is beguiling and puzzling in its simplicity. Its meaning is wide open to discussion. For example, carbon-13 levels are not set entirely by the growth of living things. A high rate of burial of organic material in the sea bed can push up the proportion of carbon-13 by removing carbon-12. If corpses of dead creatures dissolve back into the sea water and return their carbon-12, they will pull the proportion of carbon-13 down. And the carbon-13 level in sea water is also related to the prevailing abundance of carbon dioxide in the atmosphere.

There is also plenty of time for life and geography to change, during the intervals of 400 million years into which this history is divided. They are six times longer than the gulf of time since the dinosaurs died. In 400 million years the map of the continents can completely rearrange itself more than once, while the Sun and Earth pay two or three visits to different spiral arms of the Galaxy.

Variations in the variability of the biosphere in response to the changing cosmic rays may open a door to the better understanding of life's history. The emergence of animals after the climatic lurches of the last Snowball Earth episode suggests that a highly variable climate, as opposed to mere climate change, may stimulate radical innovations among living things. On the other hand, less variable conditions allow for many less radical but more elaborate refinements that produce a colourful diversity of species well suited to the prevailing climate. Too well in fact. The highly adapted creatures are more liable to succumb to subsequent changes in climate.

Another chain of interactions has been uncovered. We started this chapter's chronicle with the star formation rate and the changes in the Sun's performance. Via their effects on the influx of cosmic rays, these purely physical factors appear to have governed the Earth's climate and consequently the conditions for life. More subtly, it now seems that colder conditions are linked to bigger swings in bio-logical productivity. When he reported his findings about carbon-13 and cosmic rays, Svensmark summed up his excitement: 'If the link is confirmed, the evolution of life on Earth is strongly coupled to the evolution of the Milky Way.'

The result of the analysis should give biologists food for thought. It came from considering changes over hundreds of millions of years, which is a bit like averaging trilobites and sabre-tooth tigers. Look more closely at cosmic rays, climate and evolution over just a few million years, and you'll see a much sharper, more vivid movie of local stellar causes of climate and their dramatic consequences. So let's zoom in.

7 Children of the supernovae?

Climate change and human origins went hand in hand · They coincided with the onset of the present icy period · At least one very nearby star blew up around that time · 'Cosmic-ray winters' may have driven evolution along · Astronomers hunt for supernovae that ambushed the Earth

Did it glow near the Southern Cross, so that the bright cosmic lantern hung low over the horizon as seen from tropical Africa? Or was it high in the northern sky, vastly outshining its companions among the Seven Sisters? Astronomers will debate the point until firmer evidence comes in. But those are two candidates for the location of a nearby star that ran out of fuel and blew up more than 2 million years ago. At that time the Earth was still the planet of the apes.

The supernova must have puzzled the apes and given them restless nights. It was close enough to glow brighter than the Full Moon for weeks on end. Otherwise there was no noise or shock, nor any devastation of life by radiation sickness, which could well have happened if the explosion had been nearer. But over hundreds of thousands of years the remnant of the supernova must have sprayed the Earth with increased cosmic rays.

A different, less energetic kind of splash also arrived from that star, in the form of exotic atoms made by nuclear reactions during the stellar explosion. These were able to reach the Earth only because the supernova was no more than about 100 light-years away, making it the nearest for which any scientific record exists. Thanks to the exotic atoms, German physicists came to know of the event. And their discovery provoked discussion about the possible part played by cosmic rays in the transition from apes to humans.

Geologists, fossil-hunters and geneticists have between them traced the connection between a sharp cooling 2.75 million years ago and environmental changes favouring the first appearance of man-made tools and distinctively human genes. For anyone trying to understand his or her existence, as a thinking creature in a tumultuous Universe, it is a crucial episode. A firm explanation of that cooling event will be the jewel in the crown for climate science – and for the exploration of the Earth's cosmic neighbourhood too.

Until recently, the stars had no role to play in the hypotheses on offer for the cooling event. In the geologists' perspective, it was just one more step down a staircase of increasing cold that started 50 million years ago. Much of Antarctica was ice-covered by 14 million years ago, and soon afterwards (by geological standards of time) parts of Greenland became icy too. And global geography was changing.

Domes of high ground rose up in Africa on either side of the Great Rift Valley, and they redistributed the available rainfall, to the disadvantage of East Africa. Meanwhile India was driving into the underbelly of Asia and pushing up the Himalayas and the Tibetan plateau. That created a pool of cold in the sub-tropics. The collision of Australia with Asia helped to obstruct the routes of tropical ocean currents. And North and South America, after drifting westwards independently, were united by the completion of the isthmus of Panama around 3 million years ago. It cut the pre-existing tropical link between the Atlantic and the Pacific and rearranged the ocean currents.

Such makeovers no doubt helped to set the stage for the cooling that began 2.75 million years ago. But geographical changes occur gradually, at the speed of drifting continents, which is about as fast as your fingernails grow. And the preceding interval, starting 5 million years ago, was quite warm, with the sea 10 or 20 metres above its present level and temperatures a few degrees higher than they are now. So who switched on the freezer?

The climatic shock may well have been due to an astronomical shock. Earlier chapters have staked the claim for cosmic rays as powerful agents of climate change. They account for the broad features of our planet's climatic history over billions of years, and a far more detailed picture comes from the changes over recent millennia and decades. In between those timescales there comes this exceptionally interesting episode of climate change a few million years ago, and there's no reason why the cosmic rays should drop out of the story. On the contrary, they probably played a remarkable part.

The wanderings through the Milky Way recounted in Chapter 5 have brought the Sun and the Earth to their entry into the Orion Arm of the Galaxy. Chilling due to an increasing intensity of cosmic rays is to be expected, as these find their way here from exploded stars, tracking along the Galaxy's magnetic field. Nearby supernovae make the story particularly dramatic and quirky.

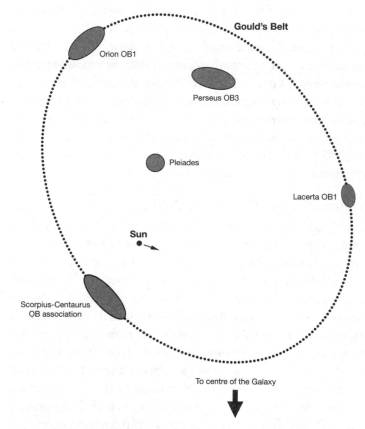

19. The ring of fire where the Sun and Earth have been hit by cosmic rays from exploding stars during the past few million years is well known to astronomers as Gould's Belt. Besides massive, short-lived stars scattered around the Belt in swarms called OB associations, the near-central Pleiades cluster may have contributed to the barrage.

Our planet's earliest encounter in the Orion Arm was with a miniature starburst – an explosive ring of fire, the remains of which we see today as an encircling string of stars called Gould's Belt. For the past few million years, the Solar System has been like a party of elderly ladies in a small boat who have blundered into the middle of the Battle of Trafalgar.

For clarity's sake, this chapter first chronicles the climatic and evolutionary events, without comment on the astronomical explanation they may demand. Then it will return to the exceptionally close supernova described at the outset. For a while it seemed to be a strong candidate as the trigger of the 2.75-million-year event, but that now looks less likely. The chapter finishes with the search in the sky for evidence of other nearby supernovae.

When the Sahel turned dusty

In the Atlantic Ocean west of Scotland, in 1995, the drilling ship JOIDES *Resolution* probed the sea-bed a kilometre down on the Rockall Plateau. When samples went to Bremen University, Karl-Heinz Baumann and Robert Huber saw clear signs of the onset of cold conditions, in a change in colour in the sediments. The arrival at the latitude of Rockall of ice-rafted debris marked the start of the present phase in the Earth's climatic history. Ever since then, ice sheets have covered substantial areas of northern lands and have often expanded in full-blown ice ages.

The alien grit first appeared on the Rockall Bank 2.75 million years ago. Sea-floor layers in this section of the core are dated rather well by a reversal in the Earth's magnetic field, when the north and south magnetic poles swapped places at a well-measured time. Earlier drillings had revealed a simultaneous sign of icing, with a jump in the proportion of heavy oxygen atoms in the sea water. Substantial ice sheets were in place in Eurasia and North America by 2.7 million years ago. So no one doubts that an emphatic change of climate occurred at that time and has been irreversible ever since.

To sample the consequences in the tropics, follow the Atlantic south from Rockall towards the Equator, and you'll find yourself off the bulge of West Africa. Seafarers passing that way are accustomed to annoying wind-blown dust that comes from the distant shore. It's a symptom of the deserts and semi-deserts that have grown during the past few million years, so that now they cover huge areas of Africa.

Notorious for its unreliable seasonal rainfall and spells of famine, the Sahel fringes the southern edge of the Sahara Desert. In the dry season, a north-easterly wind carries dust from the Sahel far out to sea. In the path of that wind, 1,500 kilometres offshore, sediments recovered by drilling in the Atlantic sea-bed in 1986 revealed when the region became seriously arid. Large amounts of wind-blown dust first appeared on that distant ocean floor approximately 2.8 million years ago. The continent was becoming drier.

At other drilling sites, closer to the West African shore and off Arabia to the east of Africa, wind-blown dust was common further back in time, because some deserts existed even when the world was warm. After the 2.8-million-year transition, the peaks of dustiness increased. And a rhythmic variation in the dust, over many thousands of years, became slower.

The sleuth who traced the drying of Africa by means of the sea-bed dust was Peter deMenocal of the Lamont-Doherty Earth Observatory near New York City. Following his first report in 1995, he continued this line of investigation by comparing his oceanic data with the fossil record of life ashore in Africa. And deMenocal made no secret of his motive: 'The reason we should care is that these results would argue climate change played a significant role in the origin of us.'

Less rain fell in Africa. Large tracts of forest shrivelled, so that the apes had a harder time finding figs. Plenty of big game appeared in the newly opened countryside, but the jaws of apes are poorly adapted to meat-eating. In effect, the change in climate provoked our earliest human predecessors to make stone tools with razor-sharp edges, for cutting the tough raw meat available on the African grasslands.

Choppers and new jaw muscles

Discoveries in 2000 and 2001 of fossil bones of ape-like creatures that trotted about on two legs around 6 million years ago threw the quest for our origins into confusion. Turning up in widely scattered places – in Kenya, Ethiopia and most unexpectedly in Chad – they sparked rival claims to the earliest ancestors of humankind. Unfortunately this international Miss Eve contest only showed that to stand up on your hind legs was no guarantee of further progress.

The various kinds of early ape-men, described with loving care by the fossil-hunters, accomplished little over millions of years. Compared with their cousins, the ordinary apes, they remained scarce. Their brains were small, their habits and diets still ape-like. If you could meet one of Nature's experimental bipeds, you might nod to it as just a leggy sort of chimp.

Censuses of animal fossils in Ethiopia's Omo Basin, which is famous for its pre-human and early human remains, show that the region was formerly covered by woodland and forest. The trees started thinning out 3.5

million years ago. When the world chilled emphatically after 2.8 million years ago, the proportion of creatures adapted to bushy grassland began to increase more significantly, so that within 400,000 years they outnumbered the woodland animals. During that interval, human beings first made their mark.

Life in general was adapting to new opportunities on Africa's expanding grasslands. Spectacular numbers of new species of antelopes attracted the big cats and other would-be meat-eaters. But apes and ape-men alike were committed by their jaws and anatomy to a mainly vegetarian diet. To eat raw meat you need either sharp teeth or a sharp cutting tool.

The oldest known products of human-like competence are stone tools that turned up in Ethiopia in the 1990s. Some are almost 2.6 million years old. In the Gona area, many are grouped together with waste fragments, as if in factories. Each finished tool was a sharp chopper made from a fist-sized cobblestone gathered from a local stream and shaped with skill of hand and eye. At Bouri nearby, unusual damage to animal bones showed the use of these tools to obtain meat and bone-marrow. The leading Ethiopian investigator of the tools, Sileshi Semaw, summed up their importance in a report in 2000.

> The beginning of the use of modified stones was a major technological breakthrough which opened windows of opportunities for effective exploitation of available food resources including high nutrient meat and bone marrow from animals. The cut-mark and bone fracture evidence from Bouri provides strong evidence for the incorporation of meat in the diet of Late Pliocene hominids as early as 2,500,000 years ago.

20. The oldest known man-made tools are choppers made from cobblestones in the Gona district of Ethiopia, almost 2.6 million years ago. They enabled early humans to eat meat, when the fruit-and-nut forests of their ape-like ancestors gave way to grassland in a cooling world. (S. Semaw)

Similar choppers, but of a later date, were associated with remains generally agreed to be human – the bones of *Homo habilis* found in Tanzania by Jonathan Leakey in 1960. The assumption is that these early people lived mainly by scavenging on other predators' kills. They co-existed around 2 million years ago with *Homo rudolfensis*, a somewhat larger person.

Why did human brains become much bigger than those of apes? Medical researchers investigating a condition of enfeebled muscles, called muscular dystrophy, reported in 2004 a genetic change that may have allowed the growth of brains to begin. A team led by Hansell Stedman of the University of Pennsylvania identified a gene that governs the thickness and strength of bite-muscle fibres controlling the jaws in all monkeys and apes. Called myh16, the gene produces a very powerful bite-muscle that completely encloses the skull and restricts its growth.

Every human being alive today has a mutated form of the gene and slighter bite-muscles. The weaker human jaws go with flatter faces, smaller teeth and rounder skulls. In Stedman's opinion, the mutation set the brain free to grow. And when the team used genetic evidence to figure out when the mutation occurred, the answer was roughly 2.4 million years ago.

The genetic dating is not exact enough to settle an argument among the fossil-hunters about who made the earliest tools. In one scenario the ape-men then living in Ethiopia, called *Australopithecus garhi*, were responsible. The genetic mutation could then take hold in them because these small-brained creatures already had the choppers that would allow them to survive with weaker jaws. Alternatively, the mutation came first and brainier people fashioned the cobbles, even though remains of *Homo habilis* or *rudolfensis* as old as the first choppers have yet to be found.

Such was the sequence of events, from the onset of northern glaciation to the bite-muscle mutation in Africa a few hundred thousand years later. It explains why the quest for the cause of the cooling became such a challenge. The discovery of an exceptionally close supernova added to the excitement and the controversy.

Flypaper for star-blown atoms

Signs of a supernova come from the bottom of the Pacific Ocean, in the form of unusually heavy atoms of iron. They are preserved as alien relics in home-grown lumps of metallic ore that litter the floor of the deep ocean. British oceanographers aboard HMS *Challenger* discovered the ferro-manganese deposits, which are sometimes flat crusts, sometimes rounded nodules. That was in the 1870s, and a hundred years later the deposits aroused excitement amid proposals to mine them from the sea-bed for sake of the manganese they contain.

In 1976 the German research ship *Valdivia* scooped up samples from the floor of the deep Pacific Ocean. At that time no one even imagined that the ferromanganese deposits might act like flypaper for catching star-blown atoms, and thus record events far out in space. But so it turned out, when in the late 1990s a team led by Gunther Korschinek of the Technological University of Munich began hunting for evidence of any nearby supernova that might have occurred in the past few million years.

An exploding star plays the alchemist on a tremendous scale. Nuclear reactions transform one element into another, creating the raw materials for planets and living things. The resulting atoms disperse and some may eventually find their way onto the Earth's surface. But severe problems faced the Munich team that wanted to identify them.

In the chasms of cosmic space, the material from the star becomes extremely sparse. Even if a supernova is fairly close, very few of the newly-made atoms will ever get here. What's more, the Earth and everything on it are made of elements acquired from similar sources – from stars that lived and died before the Sun and its family of planets came into existence. You can't distinguish a commonplace atom of iron arriving from a recent supernova from an identical atom that was here since the world began.

The trick was to find atoms made by the exploding star that have no counterparts on the planet today. They should be radioactive, with lifetimes much less than the age of the Earth, so that any examples present from the outset will have long since changed into other atoms. On the other hand, radioactive atoms that are too short-lived would expire before they reach the Earth or so soon after their arrival that present-day searchers won't find them. The search narrowed down to atoms of moderate longevity, offering a realistic hope that a nearby supernova might have occurred recently enough for some of them to have survived.

The best candidate was iron-60, a good deal heavier than ordinary iron-56 atoms. It decays radioactively at such a rate that half the iron-60 in any sample is lost in 1.5 million years. Traces will remain after more than 10 million years. Physicists calculated that a supernova was the kind of source that would most likely produce iron-60 in abundance.

To find such atoms trapped on the Earth called for exceptional skill, but at their lab in Garching by Munich, Korschinek and his team had an exceptional tool for the job. A large instrument called an accelerator mass spectrometer sorts atoms according to their weights, by pushing them to a high speed and making them swerve by means of a powerful magnet. Technical cunning minimises confusion between atoms of almost similar weights. If just one iron atom among 10,000 million million were of the peculiar iron-60 kind, the analytical system in Garching could spot it.

The news broke in October 2004 that the team had the first clear signal of a nearby supernova ever found on the Earth. The iron-60 showed up in a ferromanganese crust sample labelled 237kd, which was pleasingly flat and apparently orderly in its formation. Nearly 30 years had passed since *Valdivia* brought it aboard from a depth of nearly 5 kilometres at a station south-east of Hawaii.

The sample was not the first that Korschinek and Co. examined. In 1999, they found an apparent pulse of iron-60 atoms a few million years ago in a ferromanganese crust from another part of the Pacific. The data were sparse and the uncertainties large, with only three layers of crust distinguished. But that earlier investigation remains important as evidence that the event was registered in widely separated places, as you would expect from a supernova.

Refinements in techniques made possible a much more detailed analysis of the crust 237kd. On the ocean floor, it grew in thickness very slowly, at a rate of only 1 centimetre in 4 million years. The investigators were able to gauge the ages of 28 different layers, going back 13 million years. When they counted the atoms of iron-60, with their big mass spectrometer, they were mostly concentrated in just three adjacent layers, around 2.8 million years ago.

Until then the very existence of iron-60 atoms in the cosmos had been only a conjecture of theorists, although it found support in indirect evidence that they were formerly present in ancient meteorites. With neat timing, a NASA satellite, the Reuven Ramaty High Energy Solar Spectroscopic Imager, spotted iron-60 in the sky just as the Munich team was finding the supernova in the ferromanganese crust. Revealed by the gamma rays they emit in their radioactive decay, the iron-60 atoms are mixed with other identifiable atomic products of recent stellar explosions in the Milky Way. By 2006, the European Space Agency's INTEGRAL satellite had put the astronomical identification of iron-60 on a firmer footing.

'A cosmic-ray winter'

Astrophysicists who still try fully to understand the stars, and to figure out exactly what nuclear reactions occur in stellar explosions of various kinds, were gratified by the finds of iron-60. Among them was Brian Fields of the University of Illinois, who had speculated that nearby exploding stars ought to have left atomic marks on the Earth. He greeted the Munich results with enthusiasm.

It represents an experimental triumph and a milestone in this field. ... The iron-60 detection offers the hope that other searches for deep-ocean radioactivity can shed light on the nature of supernovae. One can turn the argument around, and use the pattern of observed radioactivities to study the ashes of the supernova's nuclear burning – probing the nuclear fires that power exploding stars.

As for the atomic detectives themselves, they were fully alert to the implications of the supernova for the Earth's inhabitants, and its possible relevance to our human origins. Gunther Korschinek and his colleagues made it their parting shot at the end of their formal report: 'This supernova could have triggered a climate change that possibly caused significant developments in hominid evolution.'

They also cited Svensmark's theory of cosmic rays, clouds and climate as a possible link. That notion had entered the speculations about a nearby supernova a few years earlier. Fields of Illinois joined John Ellis, a theorist at the CERN particle physics lab in Geneva, in suggesting that such an event could cause what they called a cosmic-ray winter. Ellis had been briefed about the possible effects of cosmic rays on clouds by another CERN physicist, Jasper Kirkby, who had heard about it from Calder and wanted support from colleagues for his proposed experiment on the subject, CLOUD. So the wheel of discovery turns.

While Korschinek and his team in Munich were trying to pin down the date of their iron-60 signal, they looked more closely into the idea of a cosmic-ray winter. They consulted Ernst Dorfi of the Institute of Astronomy in Vienna, an expert on the production of cosmic rays by supernovae. He calculated that natural particle accelerators in the expanding remnant of the exploded star would have gone on churning out cosmic rays for several hundred thousand years after the supernova event, and the influx at the Earth could have been 15 per cent higher than normal. In a public statement, the lead author of the 2004 report on the iron-60 atoms, Klaus Knie, was explicit about the possible connection.

> The accompanying cosmic-ray bombardment of the Earth's atmosphere may have caused the coeval global cooling that triggered major steps in human evolution.

The Munich supernova provoked wide interest, because its date, at around 2.8 million years ago, seemed just right for triggering the major chilling associated with the onset of ice ages 2.75 million years ago. But when the team applied a different and more accurate technique to another part of the ferromanganese crust, they arrived at a younger age. That means it was probably too late for the early icing event, although it might be linked to a later cooling spike 2.1 million years ago. Apes, ape-men and early human beings would not have failed to see the glare in the sky.

The idea of the cosmic-ray winter survived the temporary disappointment. This was certainly not the only supernova in our vicinity, and even if it were the closest it was not necessarily the most influential. A task for astronomers is to find the events recorded in the nearby cosmic scenery, starting with the question of where the Munich star itself blew up.

Candidate culprits

To deliver its iron-60 atoms from not much more than about 100 light-years away, the supernova would have been at 20 or 30 times the distance of the nearest bright star, Alpha Centauri. At present, all of the massive stars that look almost ready to explode as full-blown supernovae are further away.

One of them, at about 400 light-years, is the red giant Betelgeuse, high on one shoulder of the figure of Orion in that constellation. Its heavy mass, about fifteen times that of the Sun, means that its life will be short and its death spectacular. Already it has swollen into a huge red ball, which is a prelude to the explosion. To reflect on the meaning of light-years: if Betelgeuse goes supernova this week, our descendants won't see the glare until the 25th century.

Betelgeuse belongs to a group called the Orion OB1 association, which also includes the bright stars in Orion's Belt. Associations are clusters of stars, all born at the same time and still seen close together in the sky. OB stars are between ten and 50 times as massive as the Sun. They burn very hot, are blue in colour, and send forth intense winds. As these stars have relatively short lives – 30 to 100 million years – the OB associations are the most likely scenes of supernovae. In Orion OB1, NASA's Compton satellite saw a glow of gamma rays from aluminium-26 cooked up by stellar explosions within the past million years.

Gould's Belt is named after Benjamin Gould, an American astronomer who drew attention to it while working in Argentina in the 1870s. Its most conspicuous features are several OB associations occurring in an elliptical ring 2,400 light-years long and 1,500 light-years wide. As the Sun and its planets are right inside Gould's Belt, explosive OB stars are dotted all around us in the sky.

Chain reactions occur, in which winds and shocks from one generation of heavy stars bulldoze the thin gas that pervades the spaces between stars. The gas so compressed makes new OB stars, which will explode in their turn. In our patch of the Galaxy's Orion Arm, supernova blasts have replaced the normally cool interstellar gas by an even thinner plasma of electrified atoms, hot enough to glow with X-rays. Astronomers call this region the Local Bubble. Some prefer the Local Chimney, since they now know that the attenuated gas traverses the whole disc in which the stars of the Milky Way are concentrated. The hot plasma makes fountains into intergalactic space.

Which squadron of stars included the one that blew up closely enough to splash the Earth with that identifiable trace of iron-60 atoms? The relative positions of the Sun and its violent neighbours have changed during the past few million years. At times the action was closer than it is now. Accurate plotting of distances and motions by the European Space Agency's star-mapping satellite Hipparcos (1987–93) has helped astronomers to try to figure out where the culprit lay.

A candidate location is roughly in the direction of the Southern Cross, and therefore invisible from Europe or North America. It's the Lower Centaurus-Crux sub-group of the Scorpius-Centaurus OB association, at present around 400 light-years away. As reckoned by Jesús Maíz-Apellániz at Johns Hopkins University in Maryland, the sub-group was almost 100 light-years closer just a few million years ago. One of its outlying stars might have come within 120 light-years and then blown up.

Otherwise the gang of stars responsible may lie on the other side of the sky, in the constellation of Taurus. The Pleiades, mentioned earlier as the Seven Sisters, are the most famous cluster of them all. Although not in the ring of Gould's Belt, they may share a common origin. They are close enough for your naked eye to see several bright blue members, out of the cluster's complement of about a hundred stars. The distance to the Pleiades is increasing, so they were previously nearer still.

The largest OB stars of the cluster are no longer there,

because they have already exploded. Perhaps twenty have done so in as many million years. In Germany, Thomas Berghöfer of the Hamburg Observatory and Dieter Breitschwerdt of the Max-Planck-Institut für Extraterrestrische Physik in Garching have proposed that one of the missing Pleiades was responsible for the splash of iron-60 atoms.

Deciding the issue may have to wait for more detections of the exotic radioactive atoms, both in the sky and on the Earth. For the time being, the closeness of the Munich supernova is a continuing puzzle for astronomers. Both of those suggestions – Lower Centaurus-Crux and the Pleiades – could turn out to be wrong.

A drumbeat of stellar explosions

The hunt for more supernovae is at least as important. One or two others may have been close enough to sprinkle exotic atoms here, and those are still being sought – in ancient Antarctic ice as well as in material from the sea floor. Yet even if other supernovae in Gould's Belt were too far away to deliver atoms, they could still produce a sharp increase in cosmic rays.

The statistics of Gould's Belt suggest that a drumbeat of stellar explosions should have produced several such spikes during the past 3 million years, each capable of provoking a cosmic-ray winter of greater or lesser severity. The climatic record, from heavy oxygen counts in sea-bed microfossils, shows a series of sharp cooling events, at 2.7 million, 2.1 million, 1.3 million, 700,000 and 500,000 years ago. But to match them to supernovae will require from the astronomers not statistics, but dates.

Telescopes on the ground and in space offer several ways to identify exploded stars – the first requirement before you can figure out how old they are. The most obvious, from an astronomer's point of view, is simply to see a supernova remnant as a cloud of debris still glowing by visible and invisible light. But a harvest of some 250 objects from all around the sky takes the story back only a few thousand years.

Or you can find the remnants the hard way, by looking for radioactive atoms made in the explosions and still littering the sky. They reveal themselves by emitting gamma rays of particular energies, detectable by telescopes in orbit. For example, with NASA's Compton satellite (1991–2000), Roland Diehl of the Max-Planck-Institut für Extraterrestrische Physik and his colleagues found distinctive signs of aluminium-26 scattered all round the disc of the Milky Way, where big stars are concentrated. The gamma rays also showed hints of radioactivity in Gould's Belt, and in particular from the nearby Scorpius-Centaurus association of OB stars – now suspected as a possible source of the iron-60 atoms that reached the Earth.

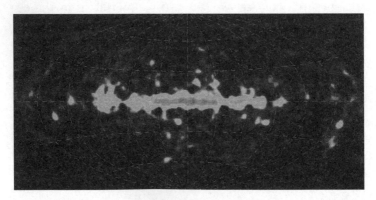

21. A whole-sky map shows emissions of gamma rays from radioactive aluminium-26 atoms created in recent supernova explosions. They completely encircle us, but are concentrated towards the centre of the Galaxy (middle of the map) and scattered around the flat disc that we see as the band of light of the Milky Way itself. The German Comptel instrument on the US Compton satellite observatory made the observations. (Roland Diehl)

Diehl's team then used Europe's INTEGRAL satellite (2002–10) to measure these gamma rays precisely enough to weigh the total aluminium-26, at three times the mass of the Sun. To produce so much of so rare an atom, a massive star must explode in the Galaxy every 50 years on average – a figure in line with the expectations of astrophysicists.

Gould's Belt stands out because it slants at an angle to the main disc. Prolonged observations by the INTEGRAL satellite should eventually capture enough gamma rays to pinpoint concentrations of aluminium-26 and other elements at particular spots in Gould's Belt, corresponding to individual supernova remnants not yet seen by other means. Then the proportions of different radioactive elements should reveal how long ago the explosions occurred.

Evidence of a third kind comes from neutron stars, the highly compressed relics of the cores of massive stars that have exploded. These were first discovered by their radio emissions, as beeping pulsars, and more than a thousand have since been found that way. Most neutron stars are probably silent at radio wavelengths, but may be detectable as pulsating sources of X-rays and gamma rays.

The prototype radio-silent neutron star is Geminga, discovered as a bright source of gamma rays in the Gemini constellation in 1972. It's about 500 light-years away. But it's rushing through the Galaxy at high speed, and Geminga may have originated in a supernova at a distance of 1,300 light-years in the neighbouring constellation, Orion, about 300,000 years ago. In the 1990s the Compton satellite logged twenty unidentified point sources of gamma rays in the direction of Gould's Belt, and these may include neutron stars. More accurate fixes are expected from NASA's GLAST satellite, flying in 2007.

To raise the hunters' spirits, a pair of objects provides

unusual evidence for a supernova of an interesting age. If a giant star that blows up has a companion star in orbit around it, the survivor flees the scene, sometimes at great speed. One such runaway is visible to the naked eye in the Ophiuchus constellation – a massive blue star called Han, or Zeta Ophiuchi. Its line of flight traces back to meet the flight-line of another runaway, this time a neutron star, pulsar J1932. Their origin was in a sub-group of the Scorpius OB2 association in Gould's Belt. You can picture a supernova blasting the neutron star from its core and also releasing its companion Han like a slingshot. Astronomers at Sterrewacht Leiden in the Netherlands estimate that the event occurred roughly 1 million years ago.

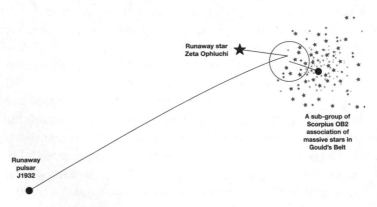

Runaway star
Zeta Ophiuchi

A sub-group of
Scorpius OB2
association of
massive stars in
Gould's Belt

Runaway
pulsar
J1932

22. *A supernova explosion in Gould's Belt about 1 million years ago set loose two runaway stars, as charted by Dutch astronomers. The pulsar J1932 is the compressed core of the exploded star, and the star Zeta Ophiuchi, otherwise known as Han, is its massive former companion. (After R. Hoogerwerf, J.H.J. de Bruijne and P.T. de Zeeuw, Sterrewacht Leiden)*

In due course, linking the various cooling events of the past few million years to bursts of cosmic rays from particular stellar explosions may not seem as difficult as it does now.

As a gamma-ray astronomer who has spent many years trying to pin down the evidence for nearby supernovae, Roland Diehl reflects on their implications for life.

Biologists talk about 'disturbance ecology', meaning the influences that storms, volcanoes and the like have, on biodiversity. Impacts of asteroids and comets are much discussed. But cosmic disturbances from stars must have often played a part in the history of life on Earth. With exactly what effect, has yet to be confirmed. To match the astronomy of the Sun's neighbourhood to the geology and the fossils will be very hard work, but we have made a start.

Few investigations have a more powerful motivation, because this one promises to illuminate the cosmic circumstances in which our human ancestors emerged. Since 1946, when Fred Hoyle in Cambridge initiated modern studies of the origin of the chemical elements, it's become a cliché to say we're made of stardust. Apart from the primordial hydrogen, every element in the human body was made in one star or another. Now there may be a more recent sense in which we owe our existence as brainy creatures to lights no longer burning in the sky. Are we the children of the climate-changing supernovae of Gould's Belt?

A necklace of new knowledge

This chapter has visited the sea floor off Africa and stars near the Southern Cross. Its exhibits have included sharpened cobblestones and a runaway pulsar. We were in pursuit of a spell of climate change spanning 400,000 years, just one ten-thousandth part of the history of life. Earlier chapters took us to an icy pass across the Swiss Alps, to the magnetic empire of the Sun, to now-distant spiral arms of the Milky Way, and to the neighbouring galaxies that per-turb our own. On an atomic scale, a laboratory experiment has revealed the chemical secrets of commonplace clouds. What we have offered is a necklace of new knowledge, threaded together by the cosmic rays.

During the 19th and 20th centuries, enquiries into the workings of Nature split apart into many narrow-minded specialities. Labels on laboratory doors, like Anthropology, Astrophysics or Atmospheric Chemistry, were assertions of independence. For critics with a traditional mind-set, the wide-ranging explorations in this book may seem ludicrous. How can the small groups of people involved in this work hope to carry any weight in so many different branches of specialist knowledge? The answer is that when teachers divided the subject-matter of science into manageable and apparently self-sufficient topics, they knew very little of the connections that Nature makes.

Time and again, expert self-sufficiency has proved to be an illusion. In the 20th century, chemistry had to meld with quantum physics, biology with crystallography, and geology with the scrutiny of planets, moons, asteroids and comets. Natural philosophers of our great-grandparents' generation would be amazed by the linkages now explored routinely in fields of research such as astrochemistry, molecular eco-logy, or the physics of brain function. Researchers at discov-ery's cutting edge now know that they have to stop acting like members of exclusive clubs, and pool their information and ideas. Forensic science has diversified from hand lenses and fingerprints to closed-circuit television and DNA test-ing. Similarly, those trying to solve the puzzles that remain about the natural world will have to combine clues of very different kinds.

From what began with his relatively simple satellite evidence that cosmic rays can influence cloudiness, Svensmark was drawn willy-nilly into new topics, from the physical chemistry of airborne sulphuric acid to the dynamics of the Galaxy, and from offbeat temperature records of Antarctica to the ever-changing productivity of the biosphere. The loop of the necklace of cosmic rays, clouds and climate is already complete, but it has plenty of room for more jewellery. To follow up the existing discoveries and their implications could already provide work for dozens of research leaders and graduate students. Some of the opportunities are outlined in the following chapter.

8 The agenda for cosmoclimatology

Energetic cosmic rays explain many details of the story · A much clearer history of our Galaxy is needed now · And a much fuller chronicle of the Earth's climate too · Our reliance on the Sun will inform the hunt for alien life · Climate science should be helpful, not grandly predictive

In the summer of 2006, with help from his son Jacob, Svensmark continued to calculate the fate of cosmic rays in the Earth's atmosphere. The German CORSIKA program enabled them to account quite precisely for the lack of any notable effect on climate of a weakening of the Earth's magnetic field. As explained in Chapter 2, the incoming cosmic rays responsible for the muons that penetrate to the lowest altitudes turn out to be so energetic that they virtually ignore any changes in the Earth's field. Only 3 per cent of the muon count is affected.

The CORSIKA calculations also shed a new light on other astronomical and solar processes that control the cosmic rays and the climate, going far beyond the question of the Earth's magnetic field. As several residual mysteries and uncertainties disappeared as if by magic, Svensmark was jubilant: 'The evidence is now massive and the story is like a fairy tale coming true.'

Cosmic rays from a nearby source, like the supernovae considered in the previous chapter, provide one example. They include a bigger proportion of energetic particles, many of which escape from the Galaxy before they get here, when they start from further away. The calculations show that a given dose of cosmic rays from a nearby supernova makes three times more muons than the corresponding dose in the general supply of galactic cosmic rays. Records of beryllium-10 and the other peculiar atoms, which are left behind by cosmic rays of relatively low energy acting at higher altitudes, will understate the climatic effect of a nearby supernova. That's the opposite of what happens when the Earth's magnetic field falters and beryllium-10 counts shoot up, with very little effect on the energetic muons involved in cloud-making and climate change.

The Sun's magnetic field is much more influential than the Earth's. Svensmark's CORSIKA calculations predict that its variations in the course of the eleven-year solar cycle should result in a 10 per cent change in the muons reaching the lowest 2 kilometres of the air. This is in line with what the muon counters show near sea level, and it accounts for the 3–4 per cent changes in cloud cover during a solar cycle.

Another mystery now cleared up was similar to the problem of the Earth's magnetic field. Every so often, the magnetised shock wave from a big eruption on the Sun can cause a sudden reduction, by 5 or 10 per cent and sometimes more, in the count of cosmic rays reaching the Earth. Called Forbush decreases after the pioneer investigator

Scott Forbush, as recounted in Chapter 2, these events could well be expected to result in a visible reduction in the Earth's cloud cover.

This doesn't happen as a rule, and the failure to find such connections served as an argument against the theory that cosmic rays influence cloud formation. The CORSIKA calculations again confirm, in this connection, that the solar shock wave affects the muon-making particles much less than the general run of cosmic rays, so no obvious effect on clouds is to be expected in Forbush decreases. Nevertheless, it happens on rare occasions. Following several events on the Sun in 1991, the Earth did indeed become a little less cloudy.

In this and other theoretical work, Svensmark was a one-man band, often labouring at home in the evening and at weekends. His small team at the Danish National Space Center was mainly occupied with the SKY experiment and the chemistry of cloud-making driven by cosmic rays. The planning, meetings and manufacturing work for the up-coming experiment in Geneva were time-consuming too. In the spring of 2006, Nigel Marsh, Svensmark's colleague of eight years' standing, had departed to Norway.

Despite these local difficulties, and perpetual worries about funding, Svensmark's sense of jubilation was heightened by the knowledge that the advances since 1996 had opened up enquiries involving scientists in other countries with very varied expertise. The subject of cosmic rays and climate had become a rapidly growing branch of science in its own right. Svensmark first named it and defined it in a proposal for the creation of a centre for research on *cosmo-climatology*.

A new field of research investigates extraterrestrial events that affect the terrestrial climate, on all timescales from fractions of a second to billions of years, and considers the consequences for life on the Earth, past, present and future.

The challenges for investigations that are now envisaged go far beyond the exchange of well-established 'textbook' information between experts in the different fields. At each turn, the subject leads directly to the frontiers of research, whether in atmospheric chemistry, astronomy, geology or the life sciences.

The molecular machinery of the clouds

Anyone out there who still believes that exploring the link between cosmic rays and clouds is a bizarre digression from orthodox meteorology and climate research, got up by a few cranks, should take note of the construction of the CLOUD facility of the European Organization for Nuclear Research (CERN) in Geneva, for which Jasper Kirkby is the spokesperson. At the time of writing, the CLOUD consortium comprises around 50 experts from seventeen institutes in Austria, Denmark, Finland, Germany, Norway, Russia, Switzerland, the UK and the USA. While we would be the last to say that weight of numbers is a reliable guide to scientific merit, an accelerator project that attracts such wide participation from recognised experts in atmospheric and solar-terrestrial physics cannot be thought frivolous. In proposing their facility, the team was upbeat in its expectations.

Space research has shown how 'big science' can make spectacular contributions to knowledge of the environment by bringing together experts from different disciplines.

After an initial replay in Geneva of Svensmark's SKY experiment from Copenhagen, the more elaborate CLOUD equipment will come on line by 2010. Using CERN's accelerated particles to simulate cosmic rays in years of experiments, it will explore the role of cosmic rays in making the specks on which clouds form, at all altitudes in the atmosphere. The team provides a cadre of committed researchers, at the dawn of cosmoclimatology.

Because they trace electric and molecular events from fractions of a second to hours and days, the CLOUD investigations fall at the rapid end of the range of timescales. But the experiments will reach also into the billion-year frame with the opportunity to test cosmic-ray action in special gas mixtures that will represent the atmosphere of ancient times, when its make-up was quite different from now.

Balloons and research aircraft carrying speck-hunting instruments will have to confirm that what the investigators see in their indoor experiments, in Copenhagen or Geneva, occurs 'for real' also in the outside air. Refining the theories of the atomic and molecular machinery of cloud-making is another task. Proof of success will come when researchers can compute the influence of cosmic rays on cloud formation and cloud properties well enough to say precisely how they contribute to present-day climate changes – globally and in all the different regions of the world, according to the variations in cosmic rays due to the Sun's changing magnetic moods.

To know the Galaxy better

Astronomers too have plenty to do to sharpen their contributions to cosmoclimatology, and not only in respect of the supernovae of Gould's Belt with which the previous chapter ended. The challenges begin with the origin of cosmic rays in the accelerators of supernova remnants. Their production is thought to become intense about 100,000 years after a star explodes.

The very sensitive ground-based array of gamma-ray telescopes called HESS, in Namibia, has detected several objects not previously known, that may be gas clouds hit by cosmic rays from elderly supernova remnants now coming 'on stream'. As announced in 2006, one bright feature lies almost in the direction of the centre of the Milky Way Galaxy. The HESS astronomers infer that the cosmic rays in that region are both more intense and more energetic than those reaching the vicinity of the Sun. Jim Hinton of the Max-Planck-Institut für Kernphysik in Heidelberg commented that the identification of the object was only the first step.

> We are of course continuing to point our telescopes at the centre of the Galaxy, and will work hard to pinpoint the exact acceleration site. I'm sure that there are further exciting discoveries to come.

X-ray telescopes in space, including Chandra (NASA, launched in 1999 and still operational) and XMM-Newton (ESA, launched in 1999 and due to continue until 2010), examine nearby and relatively young supernova remnants in great detail. Although these sources may not yet be mature cosmic-ray factories, they already show shock waves of the kind thought to accelerate particles to high energies. The satellites detect X-rays emitted by accelerated electrons, and in 2005 Chandra came up with the first convincing evidence for the acceleration of protons and other atomic nuclei.

The American spacecraft was looking at the remnant of Tycho's Star, which blew up in 1572. This was an event called Type 1A, and not one of the massive supernovae thought to be responsible for most of the cosmic rays in the Galaxy. Even so, it's thought-provoking that Chandra's astronomers found atomic matter blown out from the star travelling much faster than expected in the standard theories. Jessica Warren of Rutgers University, New Jersey, suspected that the theories will have to change.

> The most likely explanation for this behaviour is that a large fraction of the energy of the outward-moving shock wave is going into the acceleration of atomic nuclei to speeds approaching the speed of light.

The magnetic fields that thread through the Milky Way and guide the cosmic rays in our direction need to be charted better, in order to retrace the Earth's experiences of higher and lower influx of cosmic rays as it travels in and out of the spiral arms. Clues to the magnetic field come from the direction of vibration of radio waves – their polarisation, as the experts say. The radio astronomers responsible for this effort will do far better with the Square Kilometre Array. This is a global project, not yet finalised, to build a set of radio dishes in Australia or South Africa with an enormous combined area for collecting faint radio signals from the cosmos.

The concentration of cosmic rays in the flat disc of the Milky Way, which the Sun crosses and re-crosses every 32 million years, also remains uncertain. There have even been suggestions that the Galaxy's magnetic field screens us from many energetic cosmic rays produced far out in space by shock waves due to the Milky Way's motion through intergalactic space. In that case the Earth's exposure to energetic cosmic rays might increase, rather than diminish, when the Sun moves far above or below the central plane of the disc. In Svensmark's opinion, the climatic evidence points the other way, because the excursions out of the plane are linked to warmer intervals, corresponding to lower cosmic rays.

The adventures of the Sun and Earth on their travels through the Milky Way have illustrated several ways in which the influx of cosmic rays can change, with consequent effects on the climate. But knowledge of what neighbouring stars and the Galaxy in general were doing in the distant past is only sketchy so far.

Whenever the Sun runs into a relatively dense cloud of interstellar gas during its cruise through the Galaxy, the effect is to squeeze the heliosphere and the solar magnetic field that it contains. This idea provoked a suggestion about high intensities in cosmic rays recorded by beryllium-10 in Greenland and Antarctic ice around 60,000 and 33,000 years old. Encounters with small, dense gas clouds in what astronomers call the Local Fluff could have reduced the heliosphere to a quarter of its present diameter and doubled the intensity of cosmic rays coming from the Galaxy.

Priscilla Frisch of the University of Chicago has considered the influence of gas clouds on the Earth's space environment, and what she describes as 'galactic weather'. The Sun is at present in a region where the interstellar gas is unusually sparse. So reconstructions of past encounters with gas clouds should in principle form part of the analysis of the cosmic-ray history through geological time. In practice, events more than about 1 million years ago may be irretrievable. Looking to the future, other passages through the Local Fluff are entirely possible but, having studied the local map of the Galaxy, Frisch is reassuring: 'The Sun's trajectory suggests that it will probably not encounter a large, dense cloud for at least several more million years.'

The most important project of all, for the astronomical contribution to cosmoclimatology, is Europe's Gaia space mission. It's the successor to Hipparcos, which mapped the brightest stars far more accurately than ever before. Better-defined distances to the stars provided the means of gauging their ages, and so discovering the 'baby booms' in star formation that turned out to be linked to the extreme cold of the Snowball Earth episodes. But the use of a small sample of stars, and remaining uncertainties in the Hipparcos measurements, means that the star-formation story is still sketchy, with the census taken at intervals of 400 million years. What's more, the analysis so far is confined to our local domain in the Milky Way, in a region of the disc far removed from the Galaxy's centre.

Gaia will far surpass Hipparcos in the accuracy and scope of its star measurements and in its reach across the Galaxy. The multinational team aims to tell the entire story of star formation in the Milky Way over more than 10 billion years, in the central bulge, in all the different rings of the disc, and in the surrounding halo of stars. Only then will there be a clear answer to the question: 'Was star formation relatively smooth or highly episodic?' The more episodic it was, the greater will be the scope for looking for effects of high cosmic-ray influxes on the Earth.

Better knowledge of the spiral arms will be another pay-off. Hipparcos gave quite a good map of the local Orion Arm, where the Sun is now. Gaia will map all the major arms on the near side of the Galaxy, by locating the newly-formed stars that populate them. The high-precision measurements should also give a firm answer for the rate at which the spiral pattern revolves around the galactic centre, and whether the Sun's orbit is a circle or an ellipse. The results will make possible more exact calculations of when the Sun and its planets visited the spiral arms and experienced the high cosmic rays responsible for the icehouse periods in geological history.

Until Gaia, no great improvement in our knowledge of

the star formation rate during the Earth's history can be expected. The spacecraft is not due to fly until 2011 and its observations will take about five years to complete. Meanwhile, there is plenty of information from astrophysics, including observations of other spiral galaxies, for theorists to chew over. They can try to improve, for example, their grasp of the contrasts in galactic magnetism and cosmic-ray fluxes between the bright spiral arms and the darker inter-arm regions.

Another high priority is to understand better the gravitational dance of the Large and Small Magellanic Clouds, and other nearby small galaxies such as the Sagittarius Dwarf that is at present visiting the far side of our Galaxy. The aim will be to define how and when brushes with these galactic neighbours may have triggered star formation in the Milky Way. More precise calculations will need a better reckoning of invisible dark matter which adds greatly to the masses of the small galaxies. As the quest for dark matter is one of the top tasks in astrophysics, here is another example of how progress in cosmoclimatology will depend upon fundamental advances in quite different fields of research.

Puzzling rhythms on a wobbling planet

From the direct examinations of the Earth and its geological history, we have summarised evidence showing the strong influence of cosmic rays on the climate over billions of years. But it provides only a first sketch. A thorough evaluation must be aware of many other processes at work that influence the climate. They include continental growth, mountain-building, volcanic choruses, movements of continents affecting the ocean currents and circumpolar ice platforms, changes in the composition of the atmosphere, the geochemical role of life, and a long succession of impacts of comets and asteroids.

A troublesome case in point is the Milankovitch Effect, named after the Serbian engineer and climate hobbyist Milutin Milankovitch who, in the 1920s, refined previous ideas to offer an explanation of the recent ice ages. He told how the sunshine falling on different parts of the world in different seasons changes over thousands of years. The reason is that gravitational tugs from other objects in the Solar System affect the Earth's attitude in space and alter its orbit around the Sun.

Nowadays Antarctica is always covered with ice, so the critical changes are the coming and going of ice sheets on the lands of the Northern Hemisphere. And that depends (so the theory says) on whether the summer sunshine is strong enough to melt the snows of winter. Sometimes the Sun is relatively close and high in the sky during the northern summer, and it can blast the snow and ice away. But a lower Sun, further away, may leave snow lying all summer and piling up from year to year, building the ice sheets.

Astronomers can calculate the changes. The Earth's axis slowly swivels like a wobbling top, causing rhythms of around 20,000 years in the seasonal northern sunshine. It also rolls like a ship, affecting the height of the Sun in the sky, with a cycle of about 40,000 years. And over a slower cycle of about 100,000 years the shape of the orbit changes, putting the Earth nearer or further from the Sun at different seasons.

In the mid-1970s researchers detected the Milankovitch rhythms very clearly in the heavy oxygen in sea-bed cores, which is a measure of the climate. Calder was briefly involved in that adventure, in an impatient excursion into scientific research. By 1976, American and British scientists were calling the variations in the Earth's orbit 'the pace-maker of the ice ages'. Since then the Milankovitch rhythms have turned up in ancient sediments going back hundreds of millions of years, even when there were no ice ages in progress. In fact, geologists use them to calibrate their timescales. There's no doubting their reality.

On the other hand, the role of the Milankovitch Effect in the recent ice ages, where the scientific story began, has become less secure, or at least more puzzling. Most striking in the climatic record of the past million years were the switches between generally icy conditions, into relatively brief warm interludes, and back to ice again, roughly every 100,000 years. The mystery is why a quite weak effect of changes in the Earth's orbit should have such dramatic con-sequences. Somehow, it seems, the effects of the sunshine variations are greatly amplified.

Superimposed on the climate record are much briefer intervals of quick warming or cooling associated with big changes in the influx of cosmic rays. These were linked to the Sun's magnetic behaviour rather than its position in the sky. Effects of the low or high cosmic rays were much more pronounced during the generally icy phases than in the current warm interlude. This suggests that the Earth might change its sensitivity to climatic influences of any kind, whether the Milankovitch sunshine, the cosmic rays, or any other driver.

Solving the puzzle of the rhythms may therefore depend on finding the reason for a change in climate sensitivity. The very low sea level prevailing during an ice age, when huge volumes of water are bottled up as ice sheets on land, is one obvious factor. Large areas of the continental shelves are exposed, notably the North Sea, English Channel, Irish Sea and Adriatic Sea in Europe, the Bering Strait and a large area north of Siberia called Beringia, and the South China Sea. Most of the inter-island seaways of Indonesia also dry out, barring the way to important ocean currents. The sea level might make the greater sensitivity of the ice-age world to climatic influences more understandable, but would leave the quite rapid switch to warmer interglacial conditions harder to explain.

Making sense of Milankovitch has a high priority on the agenda of cosmoclimatology. One way to approach the problem will be to make simple theoretical models that capture some of the other processes at work, in addition to the changing cosmic rays. Theorists can take encouragement from the fact that ice-age episodes of the past 2 million years are exceptionally well known to geologists. That is thanks to easily accessible deposits near the top of the ocean floor, in drillable ice sheets, and in the uppermost part of the continental crust. Go deeper, and further back in time, and the view of changes of climate and their possible causes becomes much murkier.

To know the Earth better

The first known glacier of the Cretaceous Period, about 140 million years old, which put an end to arguments among geologists about whether or not there was an icehouse episode during the reign of the dinosaurs, was discovered in 2003. Firm evidence for the much older and more fearsome Snowball Earth episodes had not been forthcoming until the 1990s. That these major findings are so recent illustrates the extent of our ignorance about the Earth's climatic history. What other surprises do the ancient rocks have in store for us?

Much of the account of past climates has relied on the enormous success of ocean-floor drilling and ice-sheet drilling since the 1960s. But the oldest ocean floor is about 180 million years old and the ice cores cover a far briefer span of time. For 95 per cent of the history, since the oldest known rock formations were laid down 3,800 million years ago, everyone relies on geological exploration of the continents, which is in much worse shape.

On land, geologists are almost entirely limited to examining rocks that happen to be accessible. They may be the outcrops revealed everywhere by natural erosion, or else the local glimpses of the continental sub-surface revealed by miners, tunnellers and prospectors for oil. Little drilling is done on the continents purely for research purposes. Just as astronomers need bigger telescopes, so geologists need better opportunities to explore the crust. In 2004, a workshop on past climates organised by the US National Science Foundation called for an ambitious programme of continental drilling, modelled on the ocean-drilling experience.

For the time being, simple calculational models can attempt to make sense of what is already known, by combining the cosmic-ray estimates with knowledge of a few of the other factors at work. There are already proposals in Denmark to extend such models far beyond the past 2 million years – to the Phanerozoic Eon of the past 500 million years when links to spiral-arm crossings are best perceived, to the Proterozoic Eon when the Snowball Earth episodes occurred, and to the earliest Hadean and Archean Eons when life began under a relatively faint Sun.

It's surprising that the cosmic-ray signal comes through unambiguously in the geological data, *despite* the uncertainties. Among all the agencies that may have influenced the climate, cosmic rays are the only one so far that leaves a clear imprint over every timescale, from billions of years to months. The onus is now on those who want to contribute other elements in the story, concerning continental drift, volcanoes, impacts, ocean currents or greenhouse gases, to show how, at various times, they may have modified or countermanded the influence of the chilling stars.

Life in a violent Universe

The search for life on other worlds ranks high among the aims of researchers early in the 21st century. Besides hunting for traces of past or present living organisms on Mars, Europa and other possible targets in the Solar System, astrobiologists want to detect earth-like planets circling other stars. Both the European Space Agency and NASA have in preparation very ambitious projects for the period after 2015, to fly flotillas of telescopes in space. They will be designed to detect infrared rays coming from water vapour and other gases in the atmospheres of alien planets, which may be signs of the possible presence of life.

Philosophical questions about life in a violent Universe accompany the technical advances in spaceflight that make such projects possible. Astrophysics has revealed an endless contradiction. Living things need calm and lukewarm conditions if they are to originate and prosper, but to create and maintain a suitable environment requires events that are very dangerous to life.

The atoms in our bodies were forged in the immense heat of the Big Bang and exploding stars. They were fashioned into compounds needed for life, including water and carbon monoxide, by the action of cosmic rays at temperatures around minus 250 degrees Celsius. The Earth itself was built in high-speed collisions between asteroids, and the oceans may have come from comet ice. Even the much-reduced influx of asteroids and comets continues to cause large-scale death and destruction from time to time.

Recently, astrobiologists have confronted the role of magnetism in creating and maintaining the conditions for life on the Earth, and presumably on alien worlds as well. In 2005, the Italian astrophysicist Giovanni Bignami led the preparation of a report on the long-term space science programme, *Cosmic Vision: Space Science for Europe 2015–2025*. The report began with the need to understand the physical conditions for the emergence of life in the Universe, and it stressed the magnetic coupling between a star and its system of planets.

> The Earth's habitability, in particular, is maintained by a slowly evolving Sun that gives almost constant illumination and also screens us from energetic particles coming from supernovae in the Galaxy. The solar wind, expanding from the hot solar corona throughout the heliosphere, carries turbulent magnetic fields out to the edge of the Solar System, which drastically reduce the flux of cosmic rays. Completely to characterise the conditions needed to sustain life, especially in an evolved form, we must therefore understand the solar magnetic system, its variability, its outbursts in large solar eruptions, and the interactions between the heliosphere and the planets' magnetospheres and atmospheres.

Contributions from cosmoclimatology are therefore timely. They reveal that intense cosmic rays have overwhelmed the Sun's magnetic defences during high rates of star formation. Yet life clung on, even in the Snowball Earth events. Does our planet's particular location and environment, within the protective heliosphere provided by the Sun, contribute to its long-lasting success as an abode of life? If so, how unusual is our planet in that regard? And was the origin of life on the Earth made possible only by the *absence* of cosmic rays, due to the intense solar wind from the young Sun? Answers to these questions will help the astrobiologists to narrow down their list of targets in the search for alien life.

Another finding tells us something profound about the conditions for life – even if we are not quite sure what the message is. This is the surprisingly close link between extreme cosmic-ray intensities and extreme variability in the productivity of life, including some of the highest and lowest productivities on record, shown by the counts of carbon-13 atoms. Cosmoclimatological stress evidently has beneficial as well as harmful effects on the productivity of the biosphere. A matter for early investigation is whether the most remarkable blooms were due to a better distribution of nutrient elements, resulting from the vigorous weather systems and high rates of continental erosion associated with glacial conditions.

If bio-productivity is related to climate change, what about biodiversity, the count of all species? That is another but quite different measure of life's well-being. In the extinction of old species and the appearance of new species better adapted to an altered environment, palaeontologists have long recognised that changes of climate can help to drive evolution along. But any story of life's history told in relation to cosmic rays is complicated by impacts of comets and asteroids. These cause by far the largest loss of biodiversity, in mass extinctions, without regard to the state of the climate at the time. After such extinctions, large numbers of new species appear to fill the gaps, and bio-productivity recovers even more rapidly. This resilience of life suggests that it is in some sense pre-programmed to cope with the crises of the violent Universe.

Cosmic rays may affect the rate of evolution more

directly, by causing mutations in the genes. Was evolution more rapid when the cosmic-ray influx was particularly high? And what were the effects on the molecular clocks that evolutionists use to trace the timings of events? The clocks rely on mainly inconsequential variations in the genes that can be compared either by reading the DNA directly or by seeing small differences in the make-up of proteins that are manufactured by instructions from the genes. In addressing such comparative genomics and proteomics, cosmoclimatology meets the cutting edge of biology.

Reading the runes of the Sun

The enquiries into cosmic rays described in this book began with present-day variations in the Sun's behaviour and their contribution to current climate change. The European Space Agency has promoted wider studies of the Sun's role. Researchers from Imperial College (London), the Swedish Institute of Space Physics (Lund) and the Danish National Space Center (Copenhagen) have joined together in an ESA project called ISAC (Influence of Solar Activity Cycles on Earth's Climate). It assesses three ways in which the Sun can make itself felt: by visible and invisible light, by the solar wind impacting on the Earth's magnetic field, and by modulating the cosmic rays. The aim is to advise climate modellers on how they can include these effects in computer models simulating climate change.

In Svensmark's opinion, cosmic rays penetrating to low levels in the atmosphere have more influence on the climate than the other mechanisms suggested for solar changes; and also more effect than any other natural forcing, including volcanic explosions and the El Niño events that warm the eastern Pacific Ocean and the world as whole.

The link between cosmic rays, clouds and climate

remains just as significant today as it has been for billions of years. Any attempt to forecast the climate for the years and decades ahead will therefore rely heavily on being able to anticipate changes in the cosmic rays. Over such short timescales, when the galactic environment doesn't change noticeably, the variations in cosmic rays of climatic significance are due entirely to changes in the magnetic behaviour of the Sun. Those are what one needs to predict, to have any hope of making serious forecasts.

The responsibility falls on the shoulders of solar physicists. They are already under pressure to predict solar magnetic activity because of the dangers that sun-storms hold for astronauts and satellites in space and for power and communications systems on the ground. Anyone planning a manned mission to the Moon or Mars would like to minimise the risks by choosing a quiet period.

Predictions of sunspot counts, in their cycle of rise and fall over about eleven years, have been attempted for a long time, with only moderate success so far. In any case, the connection between sunspot counts and the frequency of solar storms is only approximate. For example, in September 2005, when the sunspot count was already low in the approach to the solar minimum, a single sunspot group exploded with nine solar flares within a week, starting with one of the most powerful flares of the past 50 years. David Hathaway of the US National Space Science and Technology Center in Huntsville, Alabama, was rueful about it: 'Solar minimum is looking strangely like solar max.'

Cosmic rays, too, conform only loosely with the sunspot count. Although generally high when sunspots are few, and reduced when there are many, there is no simple one-to-one connection. Effects on cosmic rays can lead or lag behind the rise and fall in sunspot counts by a year or so. And the influx of penetrating cosmic rays at the solar maximum around 2000 was cut to roughly the same extent as it was around 1979, when sunspots were much more numerous.

Records of cosmic rays provided by radioactive atoms suggest the presence of long cycles in solar behaviour, with the magnetic shield strengthening and weakening again over intervals of about 200 and 1,400 years. Various brave investigators have tried to read the runes of the Sun by running the cycles forward. Some say that the solar magnetic field, which more than doubled in strength during the 20th century, will be stronger still by the 2020s, implying fewer cosmic rays and clouds and a continuing rise in the global temperature. Others suspect that the field has peaked and will soon decline.

The blunt fact is that no one knows. Even the eleven-year and 22-year sunspot cycles are not fully understood. The causes of the long-term cycles are elusive, although there have been speculations about wobbles in the core of the Sun provoked by the planets orbiting around it. Big advances in solar physics, in both theory and observation, will be needed if the cosmic rays are ever to become use-fully predictable for climate forecasting.

As mentioned in Chapter 5, Eugene Parker of Chicago, the father of solar-wind research, wants the number of sun-like stars whose magnetism has been monitored routinely to be increased from ten to a thousand. That would improve the chances of detecting possible extremes of behaviour, of which we are scarcely aware. A slump of magnetic activity corresponding with the Little Ice Age on the Earth has already been seen in other stars, but not the peaks of activity that may have occurred in the Sun during the sudden warming events in the last ice age.

Would-be predictors of the Sun's future behaviour are

frustrated by the difficulty of gauging the present strength of its main magnetic field. That is because the regions around the poles are seen almost edge-on from the Earth and from most spacecraft. The Ulysses spacecraft orbits over the poles and measures the magnetic field surrounding it in space, but it does not carry the right instruments to measure the field at the surface by remote sensing. This lack will be put right by future space missions.

Europe's Solar Orbiter will use encounters with Venus over seven years to manoeuvre itself into a position from which it can see one of the poles from a slant angle of 38 degrees, compared with 7 degrees from the Earth. There is also a proposal for a Solar Polar Orbiter to be powered by solar sails and eventually to circle over the north and south poles of the Sun at half the Earth–Sun distance. That would give solar physicists a comprehensive view of solar magnetism at the visible surface for the very first time, and the hope of better prediction of the Sun's behaviour is advertised as the primary motive for such a mission.

Don't hold your breath. The Solar Orbiter is not due for launch until about 2015 and it will not have a good view of the pole until 2020. As for the Solar Polar Orbiter, it's little more than a gleam in the eye of its advocates. Although the European Space Agency has adopted it as one of the novel things to try to do in the period 2015–25, the scientists and engineers involved may count themselves lucky if their solar sailor is in orbit around the Sun by end of that period.

Meanwhile, the Sun's changing moods are too poorly understood for any forecast of solar activity and cosmic rays to be used as a basis for serious predictions of the climate changes that might be expected during the 21st century. In 2005, a Swedish prediction for the next solar cycle said it would begin in 2006 and be the weakest for 100 years. On the contrary, said an American forecast a few months later: the next cycle would start at the end of 2007 and resemble the very vigorous cycles of the 1970s and 1980s.

A wry comment on such attempts at forecasting came from three British solar mathematicians in 2006. After noting that the theories lacked solid physical underpinnings, Steven Tobias, David Hughes and Nigel Weiss offered a conjecture of their own.

> Of course it is interesting to speculate on what direction solar magnetic activity might take in the future. Recent sunspot cycles have been exceptionally vigorous … It is well known that, in the past, such episodes of high activity have tended to be followed by a dramatic crash into periods of severely reduced magnetic activity, termed Grand Minima. Although we would not presume to predict that this will happen soon, it would certainly be interesting to witness such a collapse.

To live in interesting times is commonly said to be a curse. The Grand Minima mentioned here are the Maunder Minimum of 300 years ago which coincided with the coldest phase of the Little Ice Age, and the similar solar shutdowns that caused the quite frequent freezes described in Chapter 1 which repeatedly closed the Schnidejoch pass across the Alps. Because of the uncertainties of the solar physics, cosmoclimatologists should probably be in no hurry to come to any conclusion about what will happen in the 21st century.

A constructive view of present-day climate change

As cosmic rays are a key driver of climate change, *any* attempt to offer the public a firm climate forecast for the decades ahead is scientifically rash. It can also mislead policy-makers and give the wrong advice to the people put in harm's way by rising or falling global temperatures. An early warning by Joseph Smagorinsky in the pioneering days of climate modelling by computer, in Princeton in the 1970s, still rings true: 'A bad forecast of climate is worse than no forecast at all.'

Since then the computers have become far more powerful, yet the assumptions and short-cuts used in the climate models seem ever more dubious. The possible effect of carbon dioxide on global temperatures has remained for the model-makers a matter of free-ranging estimates within wide limits. Despite repeated declarations of the need to narrow down the forecasts, predictions of warming to come during the 21st century now range from 0.5 to nearly 6 degrees Celsius, with many falling around 3 or 4 degrees. Journalists, environmental campaigners, politicians and some cheer-leading scientists discuss the consequences in a spirit of 'The end of the world is nigh'.

To correct apparent over-estimates of the effects of carbon dioxide is not to recommend a careless bonfire of the fossil fuels that produce the gas. A commonplace libel is that anyone sceptical about the impending global-warming disaster is probably in the pay of the oil companies. In fact there are compelling reasons to economise in the use of fossil fuels, which have nothing to do with the climate – to minimise unhealthy smog, to conserve the planet's limited stocks of fuel, and to keep energy prices down for the benefit of the poorer nations.

As noted in Chapter 3, cloud forcing controlled by variable cosmic rays still explains important features of current climate changes from decade to decade. A re-examination of the climatic effect of carbon dioxide is overdue. To explain why its impact often seems much less than expected has become the most urgent task in climate science.

For model-makers in the late 20th century, the benefit of focusing on carbon dioxide was that it seemed to make climate prediction a realistic goal. If you could get the numbers right, about the likely increase in carbon dioxide in the atmosphere and its effects on temperature, then you might calculate the changes in global temperature and rainfall. That the computer models were hopelessly flawed by their inability to deal properly with clouds did not by itself invalidate the ambition. But for the time being, long-term climate prediction is impossible in principle because no one can say what the Sun will do next, or how it will affect the Earth's cloudiness.

What is bad news for the would-be predictors of climate may be heartening for humanity at large. That is not just because the most alarming forecasts of global warming are likely to turn out to be exaggerated. From a better grasp of the mechanisms of climate change, more meaningful advice should become available for those in the poorer parts of the world for whom the never-ending changes of climate can spell poverty or death.

To tell someone trying to cope with a devastating flood or drought or storm that it's all the fault of global warming is no more helpful than delivering a political speech to the victim of a road accident. It contributes nothing to constructive action. As for forecasting the likelihood of floods, droughts and storms, the regional predictions from different computer models of a greenhouse-driven climate are notoriously contradictory.

Even if cosmoclimatologists attempt no long-term forecasts, they should be able to offer much better insight into the causes and patterns of regional climate change. That should help the affected populations, and perhaps enable planners to mitigate the worst consequences. The contradictory response of Antarctic temperatures is only one example of the advantage of being able to say what the surface warming or cooling should be, in each latitude band, due to a specified percentage change in cloud cover.

The Asian monsoon, powered by summer sunshine in tropical and sub-tropical latitudes and covering huge areas with a duvet of clouds, is the most important case. Billions of people depend on the monsoon rains for their prosperity. In the past, failures of the monsoon have often caused mass famine and sometimes the collapse of civilisations. Too much rain has brought unmanageable floods to India, Bangladesh and China.

In a stalagmite from a cave in southern China, the layers of annual growth enabled a team led by Yongjin Wang of Nanjing Normal University to show that, over the past 9,000 years, the Sun's activity has repeatedly affected the wetness of the rainy season. Although their report in 2005 supposed that variations in solar brightness were responsible, the data speak for themselves. A high influx of cosmic rays tends to weaken the monsoon, while low cosmic rays encourage generous rainfall. With fewer clouds over the tropical ocean, the sea surface becomes warmer and feeds extra moisture into the winds that deliver rain to the monsoon lands a few days late. Similar links between solar activity and summer rains are seen over the past 50 years, not just in Asia but also in the drought-prone Sahel of Africa.

Most explicit is K.M. Hiremath, a solar physicist at the Indian Institute of Astrophysics in Bangalore, who has examined the variations in the Indian monsoon over the past 130 years. At an international space science workshop on the solar influence on the heliosphere and the Earth's environment, held in Goa in 2006, Hiremath cited Svensmark's theory of cosmic rays and cloudiness: 'There appears to be a causal connection between the rainfall variability, the solar activity and the galactic cosmic-ray flux.'

A fascinating puzzle arises about the three-way connection between monsoons, the Sun, and the El Niño events that raise the sea temperatures in the equatorial Pacific. El Niños are sometimes followed by severe droughts in India, but not always, and use of the Pacific data in seasonal forecasting has led both to false alarms and to failures to predict the droughts. The forecasters will do better if they also take the Sun into account.

And if, as Hiremath suggests, cycles of wetter and drier monsoon seasons are linked to the 22-year cycle of solar magnetism, then you can plan accordingly. Farmers might adjust their crops, and irrigators their draw-offs, to suit the prevailing cosmic-ray intensities. And for relief agencies responsible for food aid, the advice is the same as that of Joseph to the Pharaoh. Stockpile food in the good years against the seven years of famine that will surely follow.

Climate science ought to be useful. That is the note we wish to end on, rather than on any long-term speculation about a continuing greenhouse warming or the possibility of cooling if the Sun should revert to its sulky mood of the Little Ice Age. The urge to foretell the future, before processes are fully understood, can lead scientists astray. Anyone still impatient for a climate forecast should recall the rector of Tübingen University, Johann Stoffler, whose predictive efforts were immortalised by Voltaire in *The Philosophical Dictionary*.

One of the most famous mathematicians in Europe, named Stoffler, who flourished in the 15th and 16th centuries, and who long worked at the reform of the calendar proposed at the Council of Constance, foretold a universal flood for the year 1524. This flood was to arrive in the month of February, and nothing is more plausible; for Saturn, Jupiter and Mars were then in conjunction in the sign of Pisces. All the peoples of Europe, Asia and Africa, who heard speak of the prediction, were dismayed. Everyone expected the flood, despite the rainbow. Several contemporary authors record that the inhabitants of the maritime provinces of Germany hastened to sell their lands dirt cheap to those who had most money, and who were not so credulous as they. Everyone armed himself with a boat as with an ark. A Toulouse doctor, named Auriol, had a great ark made for himself, his family and his friends; the same precautions were taken over a large part of Italy. At last the month of February arrived, and not a drop of water fell: never was month more dry, and never were the astrologers more embarrassed. Nevertheless they were not discouraged, nor neglected among us; almost all princes continued to consult them.

9 Postscript 2008 – carbon dioxide is feeble

Did a bottle of cosmic rays shape our ancestors' fortunes? · Gamma rays light up the Galaxy's explosive spiral arms · Attempts to deny the Sun's continuing role come to grief · The climatic effect of carbon dioxide is now plainly small · Global cooling would be the worst way to win the argument

After this book made its debut early in 2007, experimental work went on. The aim was to reconfirm the Copenhagen discovery of how cosmic rays help to build the sulphuric acid seeds needed for low, moist clouds to form. Not at CERN in Geneva, where lack of funding forced the CLOUD team to cancel a 2007 trial run with the accelerator. But Svensmark's group continued with their own programme by sending a small version of the Copenhagen SKY box into a deep mine in Boulby, England, to check that chemical action would cease when a kilometre of overlying rocks cut off the natural supply of cosmic rays. Discussions also began about a new experiment at the Forschungszentrum Karlsruhe, where German atmospheric physicists possessed a reaction chamber so large that the Danish team could hope to trace the seeding effect of cosmic rays all the way through to the formation of cloud droplets.

23. *Suitable for a big stride forward in cosmic-ray experiments, Germany's aerosol and cloud chamber facility* AIDA *is more than ten times larger than the reaction chamber used in Copenhagen. (Dr J. Schreiner, MPI-K Heidelberg)*

With limited time available for his theoretical work, Svensmark had difficulty in setting priorities. The agenda for new research, sketched in Chapter 8, vastly exceeded the available manpower. But he was happiest when escaping from the argy-bargy about recent global warming to the more cheerful science of the starry sky. Svensmark described his feelings in an interview for the American magazine *Discover*.

It is the potential of it that draws me on. ... I would never have thought that we would find these correlations between the cosmic rays and the evolution of the Milky Way and life on Earth. I never expected that all of these things are connected in a beautiful way.

Cosmic scenarios told how the Earth's climate and our history as living creatures marched to the drums of stellar explosions, with the chorus of cosmic rays casting their spells over our ancestors' DNA. The 'jewel in the crown', as it was called in Chapter 7, would be a cosmic-ray explanation for the cooling 2.75 million years ago that provoked the loss of African forests and the rise of tool-making and meat-eating bipeds. In a hostile review of *The Chilling Stars*, one of NASA's climate-modelling fraternity in New York City mocked this quest.

Any event even vaguely correlated to a hypothetical change in galactic cosmic rays must have been caused by it, regardless of the absence of evidence. A good example is the onset of the Quaternary ice ages 2.6 million years ago. Preliminary work suggested that a large supernova occurred at this time, and the authors discuss at length how this would have led to cooling. Unfortunately for this hypothesis, the dating of the supernova was later revised; but rather than abandon the idea, the authors simply postulate that another as-yet-undetected event must have caused the change instead.

Two developments made this scolding out of date. The team at the Technological University of Munich, who discovered the traces of a nearby supernova in material from the Pacific floor, helpfully reverted to their original date for the event, around 2.8 million years ago. That age was what first inspired the Munich team to propose a possible link between cosmic rays, the cooling climate and human evolution. They volunteered the suggestion back in 2004 (*pace* the critic) before this book was even conceived.

Secondly, it turned out that the dating of individual events might not, after all, matter very much for tracing the general climatic connections at that time. In 2007 Svensmark thought again about the predicament of the Sun and Earth during their present cruise among the explosive stars of Gould's Belt. As mentioned in Chapter 7, the stellar explosions have blown a Local Bubble of hot, thin gas that contrasts with cooler and denser interstellar gas beyond it.

The shell of the Local Bubble contains shock waves and strong magnetic fields, like a gigantic version of those at the edge of the Sun's own protective bubble, the heliosphere. As a result the shell tends to repel cosmic rays arriving from the Galaxy beyond. But it also turns back many of the cosmic rays generated by any supernova occurring locally, when they try to escape into the wider cosmos. So the Bubble is a bottle of cosmic rays. It is a chilly place for our planet to be, almost regardless of exactly when or where individual stars have blown up.

With about half a dozen giant stars dying explosively every million years, Svensmark estimated that the intensity of bottled-up cosmic rays is generally higher by perhaps 20 per cent, than in the surrounding region of the Galaxy. What matters most for the Earth's climate, in this interpretation, is the timetable of the Local Bubble's origin and growth, and how and when the Sun and Earth first encountered it. By making simple assumptions on those points, Svensmark found he could match the history of the Earth's cosmic experiences over the past 5 million years to the climatic record surprisingly well.

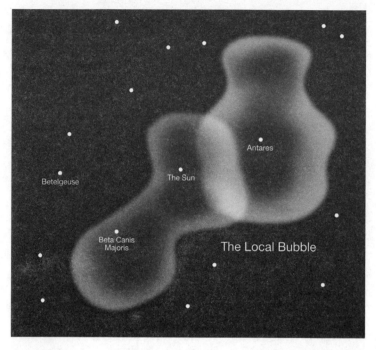

24. By bottling up cosmic rays from nearby exploding stars, the Local Bubble may be the chief governor of the Earth's present climate. (Based on a NASA illustration)

A warm spell lasting from 4.5 to 4 million years ago seemed to signal exactly when the Sun and Earth ran through the swelling shell of the Bubble. There the cosmic rays would have been fewer than either outside or inside the shell. Once inside, the Earth felt the intensifying bombardment by cosmic rays originating within the Local Bubble. The fastest cooling occurred around 2.75 million years ago in this reckoning, exactly when ice was spreading in the North Atlantic and Africa began to dry out, setting the stage for human evolution. In this perspective the

Munich supernova seems to have reinforced a general trend.

The rate of cooling then slowed down, by Svensmark's calculation, still in keeping with the geological evidence, as the climate came into equilibrium with the bottled-up cosmic radiation. That seems to be the situation our planet is in now, with the long-term icehouse conditions getting no worse. The Local Bubble has evolved into a chimney, releasing hot gas into the Galaxy's halo, and as a result the cosmic-ray count may decline in the future, and the icehouse climate relent a little.

'The good agreement with the climate history came out almost too easily', Svensmark remarked. After giving a seminar on the subject, he left the local cosmic theatre to re-examine the grander scenery of the Galaxy at large.

Charting cosmic rays in the Milky Way

Cosmoclimatology had taken a giant step forward when Nir Shaviv in Jerusalem found the link between the icy periods in the Earth's history and encounters with the Galaxy's spiral arms, as related in Chapter 5. Here was independent evidence for the role of cosmic rays in varying the climate, quite independent of Svensmark's investigations of changes related to the Sun's activity. And while the longevity of commonplace radioactive atoms, made when cosmic rays hit the planet, had previously limited the search for climate links to a few million years, in the Milky Way the historical timescale was multiplied a hundredfold.

The astronomers' description of the Galaxy remained sketchy. At most wavelengths of light or light-like radiation, the overlaps of various spiral arms obscure or confuse the human view. For half a century the radio astronomers enjoyed the special advantage of being able to chart the motions of concentrated hydrogen atoms in interstellar space. They were the first to confirm, back in the 1950s, that the Milky Way is indeed a spiral galaxy. Better radio telescopes refined the observations and adjusted the distances. But it was never easy to relate the radio patterns to the concentrations of explosive stars in spiral arms, which were responsible for the terrestrial cooling events.

Impatient for better astronomy, Svensmark had begun by turning the story around. Chapter 5 told how he used climatic data to deduce the layout of the nearest spiral arms. His report published in 2006 took advantage of the dolphin-like behaviour of the Sun, rising and diving through the flat plane of the Milky Way. That was how the Earth's temperature record weighed the Galaxy and firmed up important numbers about its dynamics.

Although Shaviv had found strong support for his inferences in the cosmic-ray exposures of ancient iron meteorites, to confirm and refine the general story called for better observational data on cosmic rays in the Milky Way. But the wild deflections by the magnetic fields of the Galaxy and the Sun mean that the direction of arrival of a present-day cosmic-ray particle says no more about its place of origin than does the glide path of an aircraft coming in to land from anywhere in the world.

One big exception concerns energetic gamma rays generated far away, when cosmic rays interact with the interstellar gas. As noted in Chapter 1, the light-like gamma rays don't feel the magnetic fields and they travel in straight lines for vast distances. When Svensmark realised that he might use them to light up the concentrations of cosmic rays in the Milky Way's explosive spiral arms, this became a new research project.

To survey the whole sky for high-energy radiation, NASA's Compton Gamma Ray Observatory, launched in

1991, carried an Energetic Gamma Ray Experiment Telescope, or EGRET. Besides spotting many individual sources, from the Moon to distant exploding galaxies, EGRET charted extensive areas of diffuse radiation all around the Milky Way, which included the gamma rays created by cosmic rays. Svensmark suspected that the most intense regions corresponded with the leading edges of the spiral arms, and especially the spaces in front of them where newly-made massive stars, cruising faster than the spiral pattern, come to their cataclysmic doom and generate the cosmic rays.

The diffuse emissions best ascribed to cosmic rays were those with a high proportion of the most energetic gamma rays. Svensmark found that they came mainly from the flat plane of the Milky Way, just as he had assumed for the cosmic rays in his dolphin study. And the energetic regions corresponded well with the directions of the various spiral arms. Most energetic of all were the diffuse gamma rays coming from the Orion Arm in which the Solar System now finds itself, with particularly good views from the Earth's vicinity in both directions along the edge of the arm.

A refinement of the galactic history of cosmic rays and climate over hundreds of millions of years was then in prospect. To pursue his interpretation of the EGRET data, Svensmark set about the task of devising a new computer model of the Galaxy, with the stars in orbit around the gravitational centre feeling extra tugs from the relatively massive spiral arms. He could also look forward to taking in fresh data one day from EGRET's successor, a new satellite called GLAST scheduled for launch by NASA in 2008. By detecting gamma rays of much higher energy, it should make the discrimination of those generated by cosmic rays all the more reliable.

As for the variable rate of star formation in the Milky Way, and the starburst events implicated in the dreadful episodes of Snowball Earth, two astronomical reports in 2007 caught your authors' attention. One was the discovery that the Magellanic Clouds may be new arrivals in our vicinity. They seem to be travelling too fast to have been in orbit around the Milky Way and therefore capable of triggering ancient starburst events in a previous close approach. An apparent match between baby booms in the Milky Way and the Small Magellanic Cloud 2,400–2,000 million years ago, noted in Chapter 6, may be a coincidence. The provocateur of that freezing starburst may have been a quite different small galaxy swallowed by the Milky Way.

Also thought-provoking, in a cosmoclimatological perspective, was the revelation that the Milky Way has had an unusually quiet history of star formation, compared with most other spiral galaxies. According to François Hammer and his colleagues at l'Observatoire de Paris, even the Andromeda Galaxy, our near neighbour in the modest Local Group, has suffered far more violent encounters, and may have had to rebuild its spiral arms more than once. Considering how lucky living things were, to scrape through the Earth's snowball episodes even in a quiet Galaxy, one might guess that most other galaxies are unsuitable for life as we know it. And even here, as Svensmark noted, the distance of the Solar System from the centre of the Milky Way may have been an important factor in our survival. Farther out or closer in, the violence of nearby star birth and death in the spiral arms would be more severe, and exposure to the chilling stars that much worse.

Comparing notes with Nir Shaviv, Svensmark found that while he as a climate physicist was looking deeper into the cosmos, for the Israeli astrophysicist it was the other way around. He was becoming more interested in the weather. They decided they might both get on better if they collaborated more closely.

'A completely different ball game'

If the planet Mars had been the subject of this book, we might have hoped for good-natured scientific sparring about the evidence, for or against a role for the Sun and the stars in changes in the weather on another world. But our concern was with the Earth, and the story that we told put a question mark beside the fashionable predictions of disastrous man-made climate change. To many climate scientists, Svensmark's nonconformity was reason enough to ignore his discoveries or else to try to stamp on them as if they were an intruding beetle.

The predictors of climatic catastrophe continued to enjoy a status reminiscent of the medieval court astrologers mentioned by Voltaire. They were often championed by eminent and influential scientists – astronomers, chemists, biologists – whose only knowledge of climate physics came from their uncritical acceptance of whatever the mainstream climatologists told them. A crescendo of political noise on the subject rose from the United Nations, the European Union and several national governments. The Norwegian parliament awarded the Nobel peace prize to the former US Vice President Al Gore for campaigning about global warming.

This was not a Nobel science prize, but Gore shared it with the Intergovernmental Panel on Climate Change. In the month that this book first appeared, that panel's scientific Working Group I issued its latest Summary for Policymakers. It refreshed the alarming predictions of a temperature rise of several degrees Celsius during the 21st

century. And it demoted the Sun to a status e\
than in a previous report in 2001. Any solar contr.
global warming was said to be only about 7 pei
carbon dioxide's effect.

The panel's scientific teams remained aloof from serious consideration of the link between solar activity, cosmic rays and clouds. But there were experts in solar physics, particle physics and atmospheric chemistry who could not plead ignorance of the theory. After all, your authors had often cooperated with them over the years. Some of them took pains to distance themselves, in the public eye, from Svensmark's politically incorrect discoveries.

One lost ally was Mike Lockwood of the Rutherford Appleton Laboratory near Oxford, who found a doubling of the Sun's interplanetary magnetic field during the 20th century, as mentioned in Chapter 3. He had been impressed by the evidence that cosmic rays affect the formation of low clouds over the ocean. But in 2007 Lockwood published a paper co-authored by another solar expert, Claus Fröhlich of the World Radiation Center in Davos, Switzerland. It was hailed around the world as conclusive proof that the Sun could not be responsible for present-day climate change, whether by way of cosmic rays or any other mechanism.

The gist of what Lockwood and Fröhlich had to say was in their title: 'Recent oppositely-directed trends in solar climate forcings and the global mean surface air temperature.' A BBC environmental reporter claimed with glee that the paper put a 'probably fatal nail' in the 'intriguing and elegant' cosmic-ray hypothesis. And he quoted a dismissive remark by Lockwood.

I do think there is a cosmic ray effect on cloud cover. … It might even have had a significant effect on pre-industrial climate. But you cannot apply it to what we're seeing now, because we're in a completely different ball game.

In the scientific paper itself, Lockwood and Fröhlich were much more generous about the Sun's past role. They noted the many solar links to climate changes over hundreds and thousands of years, as rehearsed in Chapter 1 of this book. They also allowed the Sun to make some contribution to the warming during the 20th century.

And when they went on to say that the intensification of solar activity seen during the past hundred years ended around 1985, Svensmark did not disagree with them. He might prefer to date the effective start of the present solar downturn from a minimum in the world's low-cloud cover in 1992–93, but that was a detail. So far so good.

Lockwood and Fröhlich then argued that recent trends in solar behaviour were in the wrong direction to account for what they called 'the observed rapid rise in global mean temperatures'. There they erred, because global warming

25. [Right] *Has global warming stopped? In a battle of the graphs, Mike Lockwood and Claus Fröhlich depicted surface temperatures from two official US and UK sources (top) supposedly rising rapidly in the early 21st century. Svensmark and Eigil Friis-Christensen claimed that this was an illusion created by the use of long-term running averages, over nine to thirteen years. Without such averaging, a graph of air temperatures above 1,500 metres measured by balloons (middle) showed very little increase since the early 1990s, when the Sun was beginning to calm down. The bottom graph shows ocean water temperatures down to 52.5 metres, showing a cooling since 1990. The conspicuous peak in 1998, seen in both of the lower graphs, was due to an exceptional El Niño event in the eastern Pacific. The temperatures are in degrees Celsius. (From* Proceedings of the Royal Society A *and* Scientific Report 3/2007 *of the Danish National Space Center)*

had already stopped, in response to the Sun's change of mood and in defiance of greenhouse expectations. Calder explained all this over and over again in TV interviews on the day of the paper's publication, but to little avail.

So obvious were the faults in the paper, and so busy was Svensmark with his galactic gamma rays and preparations

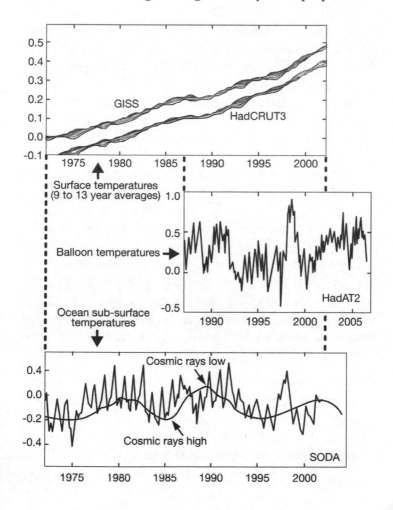

for the underground experiment in England, he hoped that some other climate physicist would do the job of formally answering them. After weeks of repeated enquiries, either hostile or friendly, he decided to join with Eigil Friis-Christensen, director of the Danish National Space Center, and rebut Lockwood and Fröhlich in print. They did so, briefly but comprehensively.

First they questioned the use of data on surface air temperature. Lockwood and Fröhlich said it did not respond to the solar cycle. Yet temperatures both in the upper air and in the ocean sub-surface water were still going up and down very clearly in response to the solar cycle. Its reported failure to show up in the surface air temperature left Svensmark and Friis-Christensen wondering about the quality of that record.

The most serious problem was with Lockwood and Fröhlich's way of displaying global warming. They used long-term averages of nine to thirteen years, which created the illusion that the temperatures were still rising rapidly early in the 21st century. One graph showed a remarkable increase of a tenth of a degree between 1998 and 2002. Yet in reality global surface temperatures had been roughly flat since 1998, and this 'apparent pause in global warming' was even plainer and of longer duration in the upper air temperatures measured by balloon. Svensmark and Friis-Christensen were confident in their conclusion.

> The continuing rapid increase in carbon dioxide concentrations during the past 10–15 years has apparently been unable to overrule the flattening of the temperature trend as a result of the Sun settling at a high, but no longer increasing, level of magnetic activity.

ball game, as Lockwood had said, was completely different, but not at all in the sense that he had in mind. The fact that the Sun was beginning to neglect its climatic duty of batting away the cosmic rays fitted all too well with what looked like the end of global warming.

After years spent doggedly on Svensmark's trail since the late 1990s, the Danish film producer Lars Oxfeldt Mortensen completed a rare piece of science history, in a TV documentary called *The Cloud Mystery*. It traced the development of the cosmic ray story, and Ján Veizer from Ottawa and Nir Shaviv from Jerusalem were among the participants. In almost the last scene to be filmed, in October 2007, Mortensen wanted to deal with the suggestion from critics that the Sun was no longer in charge.

Svensmark elected to use a graph from his paper with Friis-Christensen, which showed water temperatures in the uppermost 50 metres of the ocean, rising when cosmic rays were low, and falling when they increased. Since 1990 the data revealed an overall cooling trend. For the first time in public, Svensmark mentioned the possibility that global warming had not only paused, but perhaps gone into reverse.

Solar activity is still today regulating the temperature on Earth. It has done so in the past and will do so in the future. And it is also very interesting that the carbon dioxide is going up and the temperature is going the other way.

Squelching the carbon dioxide theory

Svensmark and Calder avoided the politics of global warming as much as possible, except when it directly interfered with the research, or with public discussion of the discoveries about cosmic rays, clouds and climate. If people wanted to economise with oil, gas and coal, in the belief that man-made carbon dioxide was driving current climate change, that was a benign consequence of a faulty idea – although a friend reminded Calder of Thomas Becket's remark in *Murder in the Cathedral*.

> The last temptation is the greatest treason:
> To do the right deed for the wrong reason.

Journalists and fellow scientists nevertheless kept asking Svensmark how he rated the contribution of carbon dioxide to climate change. While Nir Shaviv in Jerusalem, Ján Veizer in Toronto and Richard Lindzen of the Massachusetts Institute of Technology had been willing to offer evaluations, all of them on the feeble side, Svensmark still held back. As a climate physicist he just couldn't make sense of the common descriptions of molecular influences on global temperatures. He suspected that the contrary behaviour of Antarctica was just the start of new regional geography of climate change in which clouds as well as greenhouse gases play distinctive roles.

And the quest for a warming effect of greenhouse gases in the temperature records was, in Svensmark's opinion, more likely to lead the hunters to water vapour than to carbon dioxide. When the world warms for whatever reason – decreasing cosmic rays and cloudiness for example – more water evaporates and its greenhouse effect reinforces the warming. So Svensmark still declined to quantify the effect of carbon dioxide. But he suggested to Calder that the time had come to remind everyone of the strong arguments against the man-made global warming theory.

If carbon dioxide, whether natural or man-made, were an important driver of climate one would expect to see an agreement between its variations and the changing climate, on all timescales.

- Over the past 500 million years there was no correlation between climate and carbon dioxide.

- Over the past million years there was a link between carbon dioxide and temperature but the wrong way round – carbon dioxide levels followed temperature rather than leading it.

- Over the past 10,000 years there was no correlation between carbon dioxide and temperature.

- Over the past 100 years there was a rough overall link between increasing carbon dioxide and temperature.

Only the last item can be seen as possible observational evidence that carbon dioxide drives climate, and it is badly compromised by the historical details.

- Half of the 20th-century warming occurred 1905–40, when carbon dioxide levels were still quite low.

- An interlude of global cooling occurred in the 1950s and 1960s, when carbon dioxide levels were increasing.

- At the start of the 21st century global warming paused again, despite a continuing rapid increase in the carbon dioxide concentration.

- If greenhouse action by carbon dioxide drove the warming, the upper air should have warmed faster than the surface, but observations show that the opposite has been the case.

In the opinion of your authors, anyone weighing the evidence dispassionately should consider the idea of carbon dioxide as a major factor in past and current climate change to be well and truly squelched. From a scientific point of view the cosmic-ray theory is in much better shape.

- Four peaks and troughs in temperature in the past 500 million years are matched to variations in cosmic rays as observed in iron meteorites, and to the Solar System's path through the Galaxy.

- Rhythmic variations in climate over thousands of years match the variations in production of radiocarbon and other radionuclides by cosmic rays.

- Variations in the rate of warming during the past 100 years also match the variations in cosmic-ray intensity.

- The mechanism of cosmic-ray action is verified by observations of low cloud cover correlated with cosmic-ray variations, and by experimental evidence for the microphysical mechanism whereby cosmic rays accelerate the production of cloud condensation nuclei.

A little ice age? No thanks!

'They're talking about the possibility of another Maunder Minimum', Svensmark reported to Calder, on returning home from a scientific meeting in Kiruna, Sweden, in August 2007. This disconcerting reference was to the period 300 years ago when the Sun had very few sunspots and the world was cold. A discussion in Kiruna about the solar influence on climate in the past had moved on to future expectations. After scrutinising the Sun's magnetic symptoms over 400 years, Henrik Lundstedt of the Swedish Institute of Space Physics in Lund suspected that the downturn in activity around 1990 might be just the start of a more dramatic trend.

Could we be heading for another little ice age? Despite the difficulties in reading the runes of the Sun, noted in Chapter 8, the simplest betting man's argument was hard to ignore. Solar behaviour in the second half of the 20th century was exceptionally vigorous, so the post-1990 trend was more likely to continue downwards than to reverse. If global temperatures were also to follow that trajectory, the question as to whether the Sun or carbon dioxide is in charge would be answered decisively by a natural chilling event.

Neither Svensmark nor Calder wanted to win the scientific argument that way. Even if it were no worse than a return to conditions like the 1960s, global cooling would mean great hardship for humanity at large, especially among the much bigger populations of the Third World. Without quite committing themselves to a forecast (for reasons given in the last chapter), your authors were advising their friends to enjoy the global warming while it lasted.

'How can non-scientists make any sense of the competing theories being proposed, when even the observational evidence is being disputed?' An interviewer for the *London Book Review* website may have spoken for many readers when he put that question. Calder replied as follows.

Forget the politics if you can, and remember that, at the cutting edge of discovery, scientists are no more certain about what's really going on than men or women in the street. When a new finding is really surprising it falls outside the scope of existing curricula. There are neither textbooks nor highly trained people around, to be aloof in their specialist expertise. In such cases the discoverers sometimes short-circuit the academic process and take their discoveries to the general public as quickly and as directly as possible. Galileo, Darwin and Einstein all did that. They flattered their readers' intelligence as well as enlightening them, and let them make up their ow minds about whether to believe the new stories. It's in that long tradition that Henrik Svensmark and I present in plain language Henrik's astonishing realisation that our everyday clouds take their orders from the Sun and the stars. We're entirely happy that our readers, whether scientists or non-scientists, should weigh the arguments and form their own opinions, for or against us.

Sources for individuals quoted

Chapter 1

p. 13. Suter: quoted in *Die Welt*, 14 November 2005.

p. 16. Eddy: recorded interview by Spencer R. Weart, 21 April 1999.

p. 18. Parker: European Space Agency science news, 2 October 2000.

p. 20. van Geel: personal communication to Calder, 1997.

p. 21. Heinrich: personal communication to Calder, 2002.

p. 26. Bond's team: G. Bond et al., *Science*, Vol. 294, pp. 2130–6, 2001.

p. 28. Hillman: C. Hillman et al., *The Holocene*, Vol. 11, pp. 383–93, 2001.

Chapter 2

p. 33. Beer: J. Beer, *EAWAG News*, No. 58, pp. 16–18, 2005.

p. 38. Chadwick: UK Particle Physics and Astronomy Research Council press release, 4 November 2004.

p. 42. Ferrière: K.M. Ferrière, *Reviews of Modern Physics*, Vol. 73, pp. 1031–6, 2001.

p. 47. Parker: personal communication to Calder, 2000.

p. 49. Simpson: personal communication to Calder, 1994.

p. 52. NASA report on health risks on Mars: The Mars Human Precursor Science Steering Group, NASA, 2 June 2005.

p. 55. Rabi: often quoted, e.g. by S. Geer, *CERN Courier*, December 1997.

Chapter 3

p. 64. Trenberth: quoted by Jenny Hogan, NewScientist.com news service, 27 May 2004.

p. 64. Zhang: M.H. Zhang et al., *Journal of Geophysical Research*, 110, D15S02, 2005.

p. 65. Stephens: quoted in NASA press release, 15 September 2005.

p. 74.   Bolín: quoted in *Information*, Copenhagen, 19 July 1996. (In original Danish: *Jeg finder dette pars skridt videnskebeligt set yderst naivt og uansvarligt.*)

p. 74.   Kulmala: recollection by Svensmark from NOSA/ NORSAC Symposium on Aerosols, Helsingør, 1996.

p. 76.   Marsh and Svensmark on low clouds: *Physical Review Letters*, Vol. 85, pp. 5004–07, 2000.

p. 80.   Lockwood: quoted in ESA press release, 3 June 1999.

p. 81.   Marsh and Svensmark on cloud forcing: *Physical Review Letters*, ibid.

p. 82.   Intergovernmental Panel on Climate Change: *Climate Change 2001: The Scientific Basis*, Cambridge University Press, 2001.

p. 83.   Dahl-Jensen: caption for data, http://www.glaciology.gfy. ku.dk/data/ddjtemp.TXT, 1999. (In original Danish: *Ser man at Antarktis har en tendens til at 'varme op' når Grønland er 'kold' og 'køle af' når Grønland er 'varm'.*)

p. 84.   Svensmark remark about Steffensen: reporting to Calder, 2006.

p. 84.   Das: contribution by S.B. Das and R.B. Alley to Seventh Annual West Antarctic Ice Sheet Workshop, 2000.

p. 85.   Shackleton: N.J. Shackleton, *Science*, Vol. 291, pp. 58–9, 2001.

p. 88.   Pavolonis and Key: M.J. Pavolonis and J.R. Key, *Journal of Applied Meteorology*, Vol. 42, pp. 827–40, 2003.

p. 90.   Svensmark remark about Antarctic anomaly: see 'The Antarctic Climate Anomaly' under Scientific papers.

p. 91.   Blunier and Brook: T. Blunier and E.J. Brook, *Science*, Vol. 291, pp. 109–12, 2001.

p. 92.   KISS principle: see e.g. Wikipedia, http://en.wikipedia. org/wiki/KISS_Principle

p. 96.   Lamb: H.H. Lamb, *Climate: Present, Past and Future*, Vol. 2, Methuen 1977.

p. 97.   Svensmark remark about warming: reporting to Calder, 2006.

Chapter 4

p. 99. Dickens: Charles Dickens, *Bleak House*, Oxford: Oxford World Classics, 1998.

p. 100. Wallace: A.R. Wallace, *The Wonderful Century*, New York: Dodd, Mead, 1898.

p. 103. Svensmark and 'ex-president': transcript from *The Climate Conflict*, TV documentary by Lars Mortensen, Copenhagen, 2001.

p. 109. English folk tale: Joseph Jacobs, *English Fairy Tales*, London: David Nutt, 1890.

p. 111. NASA report: R.J. McNeal et al. 'The NASA Global Tropospheric Experiment', in *IGACtivities Newsletter*, No. 13, March 1998.

p. 115. CLOUD proposal: 'A study of the link between cosmic rays and clouds with a cloud chamber at the CERN PS', CERN, SPSC/P317, 24 April 2000.

p. 115. Comment circulated privately: memo received from Germany by CLOUD Consortium, 2000.

p. 118. Kirkby: personal communication to Calder, 2005.

p. 120. People in suits: Svensmark reporting to Calder, 2005.

p. 123. Svensmark on failed experiment: reporting to Calder, 2005.

p. 126. Svensmark on sparks: reporting to Calder, 2005.

p. 130. Friis-Christensen: quoted in DNSC press release embargoed until 4 October 2006.

Chapter 5

p. 133. Puggaard: C. Puggaard, translated by H.H. Howorth, *Geological Magazine*, Vol. 33, pp. 298–309, 1896 (originally in French).

p. 138. Shaviv on temperature variations: adapted from N.J. Shaviv, 'Cosmic Rays and Climate', in *PhysicaPlus*, online magazine of the Israel Physical Society, 2005.

p. 145. Shaviv on mid-Mesozoic glaciation: personal communication to Calder, 2006.

p. 146. Zhonghe: quoted by He Sheng, *China Daily*, 28 March 2003.

p. 149. Shaviv and Veizer response to Rahmstorf et al.: *Eos*, Vol. 85, p. 510, 2004.

p. 149. Royer and colleagues: D. Royer et al., *GSA Today*, March 2004, pp. 4–10.

p. 150. Wallmann: K. Wallmann, *Geochemistry Geophysics Geosystems*, Vol. 5, 2004.

p. 151. Lindzen: R.S. Lindzen, Economic Affairs, Minutes of Evidence, House of Lords, 25 January 2005, slightly edited.

p. 152. Svensmark on fossils: reporting to Calder, 2005.

Chapter 6

p. 157. Kirschvink: J.L. Kirschvink in J.W. Schopf and C. Klein (eds), *The Proterozoic Biosphere*, Cambridge University Press, 1992.

p. 160. Genzel: quoted in *Success Story*, European Space Agency Publication BR-147, April 1999.

p. 165. Shaviv and 'the long period': N.J. Shaviv, *New Astronomy*, Vol. 8, pp. 39–77, 2003.

p. 166. Fuente Marcos: R. and C. de la Fuente Marcos, *New Astronomy*, Vol. 10, pp. 53–66, 2004.

p. 169. Shaviv on standard solar models: N.J. Shaviv, *Journal of Geophysical Research*, Vol. 108 (A12), p. 1437, 2003.

p. 171. Rosing on globules: M.T. Rosing, *Science*, Vol. 283, pp. 674–6, 1999.

p. 172. Rosing on early biosphere: quoted by Paul Rincon, BBC News Online, 17 December 2003.

p. 179. Svensmark on carbon-13: H. Svensmark, 'Cosmic Rays and the Biosphere over 4 Billion Years,' *Astronomische Nachrichten*, Vol. 327, pp. 871–5, 2006.

Chapter 7

p. 186. deMenocal: flier for lecture at University of Utah, 18 February 2004.

p. 187. Semaw: S. Semaw, *Journal of Archaeological Science*, Vol. 27, pp. 1197–1214, 2000.

p. 193. Fields: First part, *Nature News*, 2 November 2004; second part, Fields' web page, November 2004.

p. 194. Knie: Technical University of Munich release, November 2004.

p. 201. Diehl: personal communication to Calder, 2006.

Chapter 8

p. 205. Svensmark on 'fairy tale': reporting to Calder, 2006.

p. 206. Svensmark on cosmoclimatology: funding application, 2006.

p. 207. CLOUD proposal, CERN/SPSC 2000–02, SPSC/P317, 1 April 2000.

p. 209. Hinton: HESS press release, 6 February 2006.

p. 211. Frisch: P.C. Frisch, *American Scientist*, Vol. 48, pp. 52–9, 2000.

p. 219. Bignami: G. Bignami et al., *Cosmic Vision: Space Science for Europe 2015–2025*, ESA BR-247, 2005.

p. 222. Hathaway: science@nasa, 15 September 2005.

p. 225. Tobias et al.: correspondence in *Nature*, Vol. 443, p. 26, 2006.

p. 226. Smagorinsky: personal communication to Calder, 1973.

p. 229. Hiremath: poster at International Living with a Star workshop, Goa, 19–24 February 2006.

p. 229. Voltaire: *The Philosophical Dictionary*, translated by H.I. Woolf, New York: Knopf, 1924.

Chapter 9

p. 233. Svensmark, 'the potential that draws me': interview in *Discover*, July 2007.

p. 233. NASA modeller, 'Clouding the issue of climate': Gavin Schmidt, *Physics World*, June 2007.

p. 236. Svensmark, 'almost too easily': message to Calder, 7 June 2007.

p. 241. Lockwood and Fröhlich: M. Lockwood and C. Fröhlich, *Proceedings of the Royal Society A*, doi:10.1098/rspa.2007.1880.

p. 241. BBC reporter and Lockwood remark: Richard Black, BBC News website, 10 July 2007.

p. 244. Svensmark and Friis-Christensen: H. Svensmark and E. Friis-Christensen, Danish National Space Center Scientific Report, 3/2007, September 2007.

p. 246. Becket: T.S. Eliot, *Murder in the Cathedral*, 1935, quoted by Gillian Spencer.

p. 248. Svensmark phone call: 1 September 2007.

p. 249. Calder interview for *London Book Review*: Pan Pantziarka, 16 July 2007; http://www.londonbookreview.com/interviews/nigelcalder.html

Scientific papers

Eigil Friis-Christensen and Knud Lassen, 'Length of the Solar Cycle: An Indicator of Solar Activity Closely Associated with Climate', *Science*, Vol. 254, pp. 698–700, 1991

Peter Ditlevsen, Henrik Svensmark and Sigfus Johnsen, 'Contrasting Atmospheric and Climate Dynamics of the Last Glacial and Holocene Periods', *Nature*, Vol. 379, pp. 810–12, 1996

Henrik Svensmark and Eigil Friis-Christensen, 'Variation of Cosmic Ray Flux and Global Cloud Coverage – a Missing Link in Solar–Climate Relationships', *Journal of Atmospheric and Solar-Terrestrial Physics*, Vol. 59, pp. 1225–32, 1997

Henrik Svensmark, 'Influence of Cosmic Rays on Earth's Climate,' *Physical Review Letters*, Vol. 81, pp. 5027–30, 1998

Nigel Marsh and Henrik Svensmark, 'Low Cloud Properties Influenced by Cosmic Rays', *Physical Review Letters*, Vol. 85, pp. 5004–07, 2000

Nigel Marsh and Henrik Svensmark, 'Cosmic Rays, Clouds, and Climate', *Space Science Review*, Vol. 94, pp. 215–30, 2000

Henrik Svensmark, 'Cosmic Rays and the Evolution of Earth's Climate During the Last 4.6 Billion Years', eprint http://arxiv.org/abs/physics/0311087, 2003

Henrik Svensmark, Jens Olaf Pepke Pedersen, Nigel Marsh, Martin Enghoff and Ulrik Uggerhøj, 'Experimental Evidence for the Role of Ions in Particle Nucleation under Atmospheric Conditions', *Proceedings of the Royal Society A*, Vol. 463, pp. 385–96, 2007 (released online 2006)

Henrik Svensmark, 'Imprint of Galactic Dynamics on Earth's Climate', *Astronomische Nachrichten*, Vol. 327, pp. 866–70, 2006

Henrik Svensmark, 'Cosmic Rays and the Biosphere over 4 Billion Years', *Astronomische Nachrichten*, Vol. 327, pp. 871–5, 2006

Henrik Svensmark, 'The Antarctic Climate Anomaly Explained by Galactic Cosmic Rays', eprint http://arxiv.org/abs/physics/0612145, 2006

Henrik Svensmark and Jacob Svensmark, 'Cosmic Ray Ionization Low in the Earth's Atmosphere and Implications for Climate', in preparation, 2007

Henrik Svensmark, 'Cosmoclimatology: a new theory emerges', *Astronomy and Geophysics*, Royal Astronomical Society, London, Vol. 48, Issue 1, 2007

Henrik Svensmark and Eigil Friis-Christensen, 'Reply to Lockwood and Fröhlich – The Persistent Role of the Sun in Climate Forcing', Danish National Space Center Scientific Report, 3/2007, September 2007

Index

Note: the emphasis is on subject matter and only some selected proper names are included. References to illustrations are in italic

A Mind of its Own

How Your Brain Distorts and Deceives

Cordelia Fine

Newly expanded for the paperback edition, this is an entertaining exposé of the amazing ways your brain tricks you in everyday life.

Can you trust your own mind? Perhaps it stumbles when faced with the thirteen times table, or persistently fails to master parallel parking. But the brain is amazing. Never before have we known so much about what it's up to. You might feel justified in thinking that you're in control.

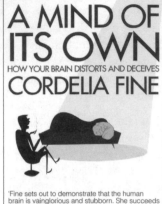

'Fine sets out to demonstrate that the human brain is vainglorious and stubborn. She succeeds brilliantly.' **Mail on Sunday**

'Fine slaps an Asbo on the hundred billion grey cells that – literally – have shifty, ruthless, self-serving minds of their own.' **The Times**

Sorry. Think again.

Your brain is fooling you. It's stubborn, emotional and deceitful. Cordelia Fine, hailed by *The Times* as a 'cognitive neuroscientist with a sharp sense of humour', takes us on an enlightening tour of the less salubrious side of human psychology. Dotted with popular explanations of the latest research and fascinating real-life examples, *A Mind of Its Own* tells you everything you always wanted to know about the brain – and plenty you probably didn't.

'Fine sets out to demonstrate that the human brain is vainglorious and stubborn. She succeeds brilliantly.' *Mail on Sunday*

'In breezy demotic, Fine offers an entertaining tour of current thinking' *Telegraph*

'This is one of the most interesting and amusing accounts of how we think we think – I think.' Alexander McCall Smith

'Consistently well-written and meticulously researched' Alain de Botton, *Sunday Times*

Paperback • UK £7.99

ISBN 10: 1-84046-798-3 • ISBN 13: 978-1840467-98-7

Why Aren't They Here?

The Question of Life on Other Worlds

Surendra Verma

Is there anybody out there? Are there other life forms lurking in outer space – or are they already here? Surendra Verma investigates ...

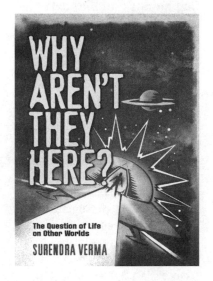

The rate of expansion of our universe is mind-blowing: imagine a pea growing to the size of the Milky Way in less time than it takes to blink. In all this infinite space that we cannot even see, let alone explore, it seems certain that there is some life on other worlds. Sir Arthur C. Clarke declared that 'the universe is full of intelligent life – it's just been too intelligent to come here'. Journalist Surendra Verma brilliantly outlines the historical, fictional, speculative and emerging scientific opinions on what alien life might be like.

From Aristotle to ET via radio, religion and reincarnation, this fast-moving narrative examines history and dispels myths before focusing on the possibilities lurking in space. In a popular and easy-to-read style, Verma uses current research to speculate what life is like on other planets, how we might communicate with it, and what Earth might seem like to visitors.

Hardback • UK £12.99 • Canada $20.00

ISBN 10: 1-84046-806-8 • ISBN 13: 978-1840468-06-9

The Cause of Mosquitoes' Sorrow

Over Two Millennia of Scientific Breakthroughs, Beginnings and Blunders

Surendra Verma

Quirky, enjoyable and entirely true instances of science going wrong, right and in totally unexpected directions from 600 BC onwards ...

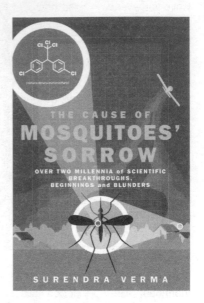

Just how did the scientific discoveries that have changed our world come about? Surendra Verma investigates the eureka moments, the serendipities and the plain errors that have peppered science's last 2,000 years. The result is a wonderfully readable insight into the mysteries of human scientific endeavour.

From the 6th century BC and Pythagoras' claims that the world was round, to the modern discovery of penicillin, Surenda Verma trawls through history in search of the more human side of science.

Who discovered anaesthesia at a party?

How did a sewage farm odour reducer benefit medicine?

Why did the Cold War prevent the West understanding heart disease?

Verma's account of philosophers, physicians, scientists and Nobel Prize winners is a highly informative and a brilliantly light-hearted account of how often Lady Luck can play a part in the scientific process.

Paperback • UK £7.99 • Canada $16.00

ISBN-10: 1-84046-831-9 • ISBN-13: 978-1840468-31-1

The Science of Doctor Who

Paul Parsons

Have you ever wondered how Daleks climb stairs? Or where the toilets are on the Tardis? This hugely acclaimed guide tells all ...

Since 1963, the journeys of the Time Lord have shown us alien worlds, strange life forms, futuristic technology and mind-bending cosmic phenomena. Viewers cowered terrified of Daleks, were amazed with time travel, and travelled through black holes into other universes and new dimensions.

The breadth and imagination of *Doctor Who* have made the show a monumental science fiction success. *BBC Focus* editor Paul Parsons explains the scientific reality.

Discover:

- why time travel isn't ruled out by the laws of physics

- the real K-9 – the robot assistant for space travellers built by NASA

- how genetic engineering is being used to breed Dalek-like designer life forms

- why before long we could all be regenerating like a Time Lord

- the medical truth about the Doctor's two hearts, and the real creature with five.

Paperback UK £8.99 • Canada $18.00

ISBN 10: 1-84046-791-6 • ISBN 13: 978-1840467-91-8

Atom

Piers Bizony

The official tie-in to a
brilliant new BBC TV
series, *Atom* is the story
of the greatest human
scientific discovery ever.
No one ever expected the
atom to be as bizarre, as
capricious, and as weird
as it turned out to be. Its
story is one riddled with
jealousy, rivalry, missed
opportunities and
moments of genius.

John Dalton gave us the
first picture of the atom in
the early 1800s. Almost 100 years later came one of the most
important experiments in scientific history, by the young misfit New
Zealander, Ernest Rutherford. He showed that the atom consisted
mostly of space, and in doing so turned 200 years of classical
physics on its head.

It was a brilliant Dane, Niels Bohr, who made the next great leap –
into the incredible world of quantum theory. Yet he and a handful
of other revolutionary young scientific Turks weren't prepared for
the shocks Nature had up her sleeve. Mind-bending discoveries
about the atom were destined to upset everything we thought we
knew about reality. Even today, as we peer deeper and deeper into
the atom, it throws back as many questions at us as it does
answers.

Hardback • UK £12.99 • Canada $20.00

ISBN 10: 1-84046-800-9 • ISBN 13: 978-1840468-00-7